Praise for *Fibershed*

"*Fibershed* is a story of vision, persistence, and kindness. With patience and grace, Rebecca has restored a sense of gratitude for the overlooked grasses and herbaceous plants that were once our second skin. From the living world around her, she has stitched together the broken strands of textile arts, creating an economy of place where makers are artists and clothing is revered."

—**PAUL HAWKEN**, author of
Blessed Unrest; editor of *Drawdown*

"Rebecca Burgess is the Alice Waters of the slow fiber movement. Within the pages of *Fibershed*, she proves that carefully clothing oneself is a revolutionary act. While many wait for distant corporations and governments to curb toxic, unethical, and extractive industrial practices, Burgess demonstrates that the revolution is at hand in our own backyards."

—**DAN MALLOY**, surfing ambassador,
Patagonia; cofounder, Poco Farm, Ojai, CA

"The sins of oil-based fibers are well known, but lesser known are those of plant- and animal-based fiber production—themselves major contributors to global desertification and climate change. If we want to offer hope to future generations, we will have to root not only the food we eat, but the clothing we wear in a new, regenerative agriculture that manages livestock using the holistic planned grazing process. Rebecca Burgess's well-researched book stokes a fire that has already been lit by many organizations collaborating and networking around the globe, and connects the dots between our clothing and our life-supporting environment. I would encourage everyone who wears clothes and has any concern for future generations to read this book."

—**ALLAN SAVORY**, president and
cofounder, Savory Institute

"Rebecca has made an incredible contribution to the slow fashion movement through her organizing and advocacy work with the Fibershed organization. I'm thrilled to know that this work is now available to a broader audience through this thoughtful book. May we all learn from her wisdom, research, and knowledge as we create even deeper connections between farms, fiber art, and fashion."

—**KATRINA RODABAUGH**,
author of *Mending Matters*

"This is an important book. It is bold, practical, optimistic—a vision of how things must be."

—**KATE FLETCHER**, professor,
Centre for Sustainable Fashion,
University of the Arts, London, UK

"*Fibershed* is a deeply informed exploration of the political ecology of clothing and an urgent invitation to a new way of being in the world; one that respects the soil, the cycles of the year, and life itself. In this visionary manifesto of hope, Rebecca Burgess chronicles a personal journey with profound global implications: Human economies need not result in the degradation of either human culture nor the environment, but might, if done well, lead to the enrichment of both."

—**JEFFREY CREQUE**, PhD,
Director of Rangeland and Agroecosystem
Management, Carbon Cycle Institute

"*Fibershed* is a must-read for all clothing brands, whether years into their sustainability journey or just at the beginning. Burgess encourages us to think deeply and holistically about the impacts of fashion, reconsider our industry's model of overconsumption, and to approach flashy biotech solutions with a critical eye. *Fibershed* proves that fashion can be a force for good, empowering farmers and makers while supporting local communities with Climate Beneficial textile supply chains."

—**MEGAN MEIKLEJOHN**,
Sustainable Materials and Transparency
Manager, Eileen Fisher

Alpacas raised by Sandy Wallace above
the Nicasio Reservoir in Northern California.

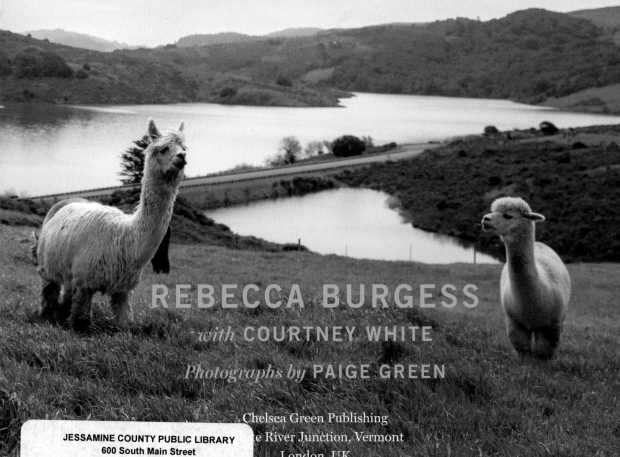

FIBERSHED

Growing a Movement of Farmers, Fashion Activists, and Makers for a New Textile Economy

REBECCA BURGESS

with **COURTNEY WHITE**

Photographs by **PAIGE GREEN**

Chelsea Green Publishing
White River Junction, Vermont
London, UK

Unless otherwise noted, all photographs copyright © 2019 by Paige Green.

Front cover photograph by Paige Green of Geana Sieburger of GDS Cloth Goods (www.gdsclothgoods.com)
holding handwoven blankets by Meridian Jacobs (www.meridianjacobs.com).

Fibershed icons by Andrew Plotsy, courtesy of Fibershed.

Editor: Makenna Goodman
Project Manager: Patricia Stone
Copy Editor: Laura Jorstad
Proofreader: Ellen Bingham
Indexer: Shana Milkie
Designer: Melissa Jacobson

Printed in the United States of America.
First printing October 2019.
10 9 8 7 6 5 4 3 2 1   19 20 21 22 23

Our Commitment to Green Publishing
Chelsea Green sees publishing as a tool for cultural change and ecological stewardship. We strive to align our
book manufacturing practices with our editorial mission and to reduce the impact of our business enterprise in
the environment. We print our books and catalogs on chlorine-free recycled paper, using vegetable-based inks
whenever possible. This book may cost slightly more because it was printed on paper from responsibly managed
forests, and we hope you'll agree that it's worth it. *Fibershed* was printed on paper supplied by Versa Press that
is certified by the Forest Stewardship Council.

Library of Congress Cataloging-in-Publication Data
Names: Burgess, Rebecca, 1977– author. | White, Courtney, 1960– author.
Title: Fibershed : growing a movement of farmers, fashion activists, and makers for a new textile economy /
    Rebecca Burgess, Courtney White.
Description: White River Junction, Vermont : Chelsea Green Publishing, 2019. | Includes bibliographical
    references and index.
Identifiers: LCCN 2019027521 (print) | LCCN 2019027522 (ebook) | ISBN 9781603586634 (paperback)
    | ISBN 9781603586627 (ebook)
Subjects: LCSH: Textile fiber industry—Environmental aspects. | Plant fibers—Environmental aspects.
    | Animal fibers—Environmental aspects.
Classification: LCC TS1540 (print) | LCC TS1540 (ebook) | DDC 677—dc23
LC record available at https://lccn.loc.gov/2019027521
LC ebook record available at https://lccn.loc.gov/2019027522

Chelsea Green Publishing
85 North Main Street, Suite 120
White River Junction, VT 05001
(802) 295-6300
www.chelseagreen.com

Contents

Japanese indigo seedlings.

Introduction

On an early morning in July in 2010, I stood at the northern edge of the San Francisco Bay looking out over a stretch of fog filtering through the Golden Gate Bridge. I was waiting for a group of children who were part of an ecological arts class I had developed for our local science museum. Within moments I heard the sound of small feet scampering across cement floors, their voices echoing through the high-ceilinged World War Two–era hallways. This was the day we made our annual field trip to the old fortifications to observe the Mission blue butterfly habitat that was being revived among the remains of the war bunkers. I usually looked forward to this walk together, but that morning I was overwhelmed with concern.

A few moments earlier I had called a company in Santa Cruz that focused on growing specialty plant starts to see how their experiments with my indigo seed were doing. For several years I had been growing organic indigo to produce a natural blue dye for my textile creation work. By 2010 the project had expanded to the small farm scale, and there was no longer enough space in my apartment for the seedling transplant trays. To implement this scaling-up efficiently, I leased a small plot of farmland twenty minutes away. I also decided to test the germination rate and the compatibility of my seeds with mechanized plant-start equipment. The company in Santa Cruz offered the kind of help I needed for expanding the project while allowing me to concentrate on my paid work as an environmental educator. On the call that morning, I learned that the company's second germination test had gone well—so well, in fact, that the company had already shipped the indigo plants out—except they had no idea which farm they had been sent to.

I knew the tiny seedlings in their one-inch-deep containers weren't going to live long without being watered, and the recent history of my efforts ran through my mind as I imagined the plants being delivered to an unknown location and withering rapidly. Four years of seed saving, countless days of seed cleaning, and my dream of creating a completely homegrown fermented indigo dye for vat production would all be lost if those seedlings went without water for even a few hours.

Fortunately I received a call a short while later from color-grown cottonseed breeder Sally Fox, a colleague who lived approximately two hours northeast of me in the Capay Valley, near Sacramento. Sally had received a message that morning from a neighboring organic farm. The caller told her that an unidentified set of plant trays had been delivered to their site along with their tomato and bell pepper seedlings. He suspected the plants were part of Sally's plant breeding projects, and Sally thought that they were likely part of my indigo project (how this all worked out so well, I'll never quite understand, but I'm eternally grateful to them). I called the farm owner, Tim Mueller, who expressed an interest in both the plants and the project and told me he was happy to keep the remaining indigo plants watered in his greenhouse. What did he mean, remaining plants? The rest, he said, had already been planted. I was very relieved to hear the crop had found its way to the soil, even if that soil was two and a half hours from my home.

That weekend I visited Riverdog Farm and was heartened to see an integrated crop-and-livestock system at work. Chicken tractors dotted the fields, pomegranate hedgerows lined the dirt roads, and a large peach tree hovered alongside the office. After I arrived, Tim jumped into my car, and we drove down the road to leased acreage where rows of bell peppers had recently been planted. There among the food crops were two long rows of indigo, totaling six thousand plants. I'd never seen indigo grown like this before. It didn't look like my garden, of course, where I had begun the indigo project years ago, but it didn't resemble the small farm I was managing, either. These meticulously planted, 150-foot-long rows were a hopeful sight.

Standing there, looking at the indigo in its new home, I was filled with the inquiry that Michael Pollan expressed in his book *Botany of Desire*, a question that arises from looking more closely at the relationships between humans and the species that we cultivate and help to multiply: "Who is domesticating whom?" The rich and nontoxic shade of blue that *Persicaria tinctoria* yields has motivated energetic human support over the centuries to expand the terrain where this crop was originally cultivated. By working its magic on humans, the plant has spread its genetics far and wide—as I was now helping to do. As a result of Riverdog Farm's efforts, I was now assured of having enough material to create the critical mass of dried leaf—440 pounds—needed for composting in order to make what is known in Japan as *sukumo*, a naturally

The Grow Your Jeans project involved hours of dedication by team members. Rebecca Burgess grew and processed the indigo used to dye Sally Fox's organic cotton yarn, *shown here*.

grown leaf concentrate of temperate-climate indigo that can be used year-round for indigo dyeing work.

For the rest of the summer, I happily harvested and processed the crop with my friends, and for the next two years, we grew indigo at Riverdog Farm, successfully achieving the critical mass of dried leaf we needed from each harvest. We were able to build a stock of composted indigo leaf that has now been used by the Berkeley Art Museum, in an experimental program at clothes maker Levi Strauss, in a prototype project for Fibershed called Grow Your Jeans, as well as in a range of dye workshops that have been offered to the local public. The indigo project is now in its seventh year and continues to grow, even in the face of California's drought conditions, to include a number of new farmers and a new generation of *Persicaria tinctoria* stewards.

Taking on the responsibility for introducing (or reintroducing) a new crop like indigo to a community can be a daunting task. But there is

Leslie Terzian of TangleBlue wove the cloth for the Grow Your Jeans Project.

a deep, almost cellular response in humans when we take on this type of work, one that creates an unbreakable bond. In fact, it has been an incredible surprise to see how many people are similarly committed to the cause of regionalizing and relearning what it means to produce your fiber and dye. Through my work I have seen the act of growing our own clothes to be a galvanizing community experience.

Because we have been disconnected from the impacts our clothes have on land, air, water, labor, and our own human health for such a long time, we've been lulled into a passive, non-questioning state of being as consumers. When we begin reconnecting these dots, however, we create opportunities to build new relationships that are rooted in sharing skills, physical labor, and creativity, all of which carry meaning, purpose, and a way to belong to one another and to the land. While there has been important work in recent decades to ensure access to safe, local, nutritious food as a culture, we have largely overlooked the production of fibers and dyes that make up our clothing. In fact, when people hear the word *clothing*, most automatically think, *Oh, I don't care about fashion*, and assume it has nothing to do with them. But clothing—like food—matters because we directly engage with it every single day. Clothing is a multifaceted industry that involves many of the same supply-chain dynamics as the food industry, starting with its roots in agriculture and dependence upon the land.

Models from the Grow Your Jeans fashion show pose for a group photo.
Left to right: Celeste Thompson, Leslie Channel, Thyme Francis,
Dario Slavazza, Alycia Lang, Sally Fox, Sophia Zuchowski.

What Do Clothes Have to Do with Agriculture?

The simple answer to this question is: a lot. On average, over 80 percent of the cotton grown in the United States annually is genetically modified to withstand the use of a range of herbicides and pesticides, and less than 1 percent is certified organic. And while two-thirds of Americans support GMO labeling for their food, few understand the role GMOs play in their clothing. In fact, we have yet to broach any large-scale public discussion of how GMO agriculture as a whole is impacting the health and diversity of our landscapes, rural economies, and personal health. Due to the omission of these larger conversations we've largely left the genetic engineering of fibers out of the land-use ethics debate altogether, and as a result there is little to no transparency offered on garment hangtags enabling us to determine if our clothing is genetically modified or not. Unless we are searching out and purchasing Global Organic Textile Standard (GOTS) certified garments. As a result of the large gap between our knowledge of how clothing is made and where the ingredients are sourced from, when we make decisions as a consumer on what to buy, we are largely making them blindly.

Consider this, for example: American-made wool garments are rare despite the United States being the fifth-largest wool-producing nation in the world. Almost all of our wool socks and suits are made in Australia, New Zealand, and China. Beyond that, over 70 percent of the fibers we wear originate from fossil carbon, and almost every garment is colored with dyes that are sourced from fossil carbon. Plastic microfibers that are introduced into rivers, streams, and oceans as a result of the washing of synthetic clothing are contaminating the marine food web as well as our drinking water. Significant concentrations of fiber lint have been found in the deepest ocean habitats with yet-to-be-determined consequences. Working conditions for textile employees are notoriously challenging, and less than 1 percent of clothing sold in the United States is Fair Trade Certified. And now extreme genetic engineering is being offered to consumers as a high-tech solution to the issues created by our antiquated, synthetic, toxic chemistry; fossil carbon dependencies; and overconsumption. Most wearers have no idea that these proprietary biotech technologies share a host of supply-chain and business architecture problems and have not yet been assessed for their potential negative consequences to land, water, flora and fauna, and regional economies despite any claim they might make to the contrary.

Improving the existing centralized systems of textile production, currently based largely overseas in countries with minimal attention to human rights and weak environmental standards, is one avenue for social and environmental change that offers rays of hope. But it has not been without countless disappointments. And novel technologies also have a role to play in reducing negative impacts of the garment industry. But both of these tools for reform *on their own* do nothing to transform the existing power dynamics and economic models that provoked the environmental and labor rights catastrophes we are currently digging ourselves out of globally. And yet it is these two strategies that dominate the agendas of sustainability teams at the world's largest textile companies, that are written about and debated within the trade group journals, and that receive awards at global textiles conferences, reaping investor capital. As a result the conversation that inserts economic and climate justice into the DNA of the systems-change thought is still waiting for its day in the sun.

This book seeks to open the door for that conversation, while recognizing that many more individuals and organizations are also expanding this dialogue on a daily basis. In the following pages you will read a vision of change that focuses on transforming our fiber and dye systems from the soil up. This vision embraces everyone involved in the process, including farmers, ranchers, grassroots organizers, designers, manufacturers, cut-and-sew talent, crafters, fashion pundits, investors, transnational brands, and you—the wearer. It is a vision for globally impactful solutions that consider and provide a voice on how to reconfigure the seat of power and begin putting decision making into the hands of those most familiar with the social and ecological infrastructure of their communities. It is a vision that enhances social, economic, and political opportunities for communities to define and create their fiber and dye systems and redesign the global textile process. It is place-based textile sovereignty, which aims to include rather than exclude all the people, plants, animals, and cultural practices that compose and define a specific geography.

I call this place-based textile system a *fibershed*. Similar to a local watershed or a foodshed, a fibershed is focused on the *source* of the raw material, the *transparency* with which it is converted into clothing, and the *connectivity* among all parts, from soil to skin and back to soil. In the fibershed where I live, for example, natural plant dyes and fibers such as flax, wool, cotton, hemp, and indigo are being

> I call this place-based textile system a *fibershed*.

grown using practices that are both traditional and modern, and many of these cropping and livestock systems are showing benefits that we are just beginning to document in detail, such as ameliorating the causes of climate change, increasing resilience to drought, and rebuilding local economies.

Leslie Channel beside a bed of Japanese indigo while wearing organic and biodynamic jeans that were grown, woven, and sewn in the Northern California Fibershed. Sally Fox's color grown cotton yarn was dyed in Rebecca Burgess's fermentation indigo vat, then woven into fabric by Leslie Terzian, which Daniel DiSanto made into bespoke jeans that he designed. The pokeberry top was dyed and knit by Monica Paz Soldan from Hazel Flett and Sue Reuser's Cormo and Romeldale cross wool yarns.

Fibershed systems borrow considerable inspiration and framework design from the Slow Food movement, which can be traced back to 1986 when the movement's founder, Italian farmer Carlo Petrini, organized a protest against the opening of a McDonald's chain restaurant near the Spanish Steps in Rome. Petrini's galvanizing quote ushered in global affirmation of the need to attend to our food system: "A firm defense of quiet material pleasure is the only way to oppose the universal folly of Fast Life." The Slow Food movement quickly gained a following, attracting rural and urban residents alike. It joined an energetic effort by people around the world to address how our food is farmed, who is farming it, how it is processed, and who has access to it. Today these questions guide the mission statements of thousands of non-governmental organizations (NGOs) focused on reforming our food system, and yet we do not see an equally formidable NGO presence that has developed a strategy to support a separate but no less significant product from our working landscapes: our clothing. But there is a grassroots movement afoot to change this, led by farmers, ranchers, artisans, and small- to mid-scale textile manufacturers. Biosphere-based fibers such as flax, nettle, hemp, wool, milkweed, cashmere, angora, and cotton are making a remarkable comeback, and awareness is being raised on the undeniable fact that the soil that feeds us is also the soil that clothes us.

This is the story of that resurgence.

Entering My Fibershed

Fibersheds are dynamic, evolving living systems that are far more complex than I can begin to comprehend. They exist on their own accord, and the narrative in this book explores my journey into this living system and also highlights regional textile projects that have emerged in affiliated communities. Fibersheds are places where thousands of generations of people have lived and thrived prior to my existence. It is my intention, as well as that of many others, to help ensure that future generations will continue to have an opportunity to flourish in relationship with their place-based textile cultures.

In 2010 I began a personal journey within my home fibershed—a 150-mile radius from my front door in San Geronimo in Northern California—by challenging myself to create and wear a wardrobe for one year made from locally grown fibers and natural dyes and produced by local labor, including myself, new and old friends, and family. This "wardrobe challenge year" brought me into a direct relationship with the land in the form of farms, ranches, and open spaces. I refined my own textile-making skills and created close bonds with the artisans in my community who could turn plant and animal fibers into beautiful garments. Coming into direct and consistent physical contact with the biological source of my clothing transformed my understanding and appreciation for my fibershed.

My journey began with an investigation into plant-based colors. Developing a natural dye practice set within a defined fibershed honed my relationship to the species that existed within the boundaries of my home geography. During the one-year wardrobe challenge, my plant palette was derived from a host of endemic (native) and non-native plant species that were part of my own garden or were sourced from friends' gardens and wild hillsides. The chaparral, oak woodland, and coastal shrub plant communities produce a range of colors, including greens, orangey pinks, and browns from coffeeberry, toyon, and hinsii walnuts. Non-native species in our region provided pinks from pokeberry and yellow from weld.

> My journey began with an investigation into plant-based colors.

The one color that was difficult to achieve locally (at the time) was blue. Synthetic indigo dye has been widely available in this country beginning in 1876, and jeans have been colored with fossil-carbon-derived aniline dyes ever since. I decided to try to grow my own blue. However, unlike most natural dyes, indigo is not soluble in water. To create a usable dye, the pigment must be extracted from the leaves and combined with materials that will make the color accessible to fiber. And most dyers, including myself, originally purchased pigment from other countries to ferment in our own indigo vats. However, this imported indigo came from the tropical species *Indigofera tinctoria* and would not grow in my region of Northern California despite many attempts. For several years I experimented in my garden with an annual blue-producing species known as *Persicaria tinctoria*. I had found ways to extract its blue in the summer with fresh leaf material, but I had not yet mastered methods for extracting enough blue pigment to ferment for year-round dye projects from the plants I was growing. Blue pigment is approximately 2 percent of the weight of the *P. tinctoria* plant, and growing enough to make pigment to produce my own fermentation vat was a discovery process.

After some research I came across the work of Rowland Ricketts, who had spent time apprenticing in Japan, farming indigo and composting the plant's dried leaf to produce a material known as sukumo. The fermentation recipe that Ricketts used was based on homemade hardwood ash lye, wheat bran, and lime—all readily available within my region's raw material resource base. I reached out to Ricketts to learn more about the requirements for making compost from the *Persicaria* dried leaf and discovered that you need a critical mass of leaf material to generate enough heat for the thermophilic bacteria in your compost pile to thrive. Thermophilic bacteria need and require heat to reproduce, and with the right balance of bacteria, these small life-forms are able to eat the cellulose of the *Persicaria* leaves and leave behind the blue pigment in a material that looks like dark blue-black compost. The critical mass of dried leaf required for composting sukumo is in the range of 440 pounds, and it must be composted on a uniquely constructed floor made of rock, sand, rice hulls, and compacted clay for one hundred days, turned every seven days with a gentle spray of water.

Ricketts came out and helped us with the construction of the composting floor, working with a team of local artisans. We had the rice hulls ready, sourced from a California rice-drying facility, bagged clay from our local clay mine, added sand, and we were ready to go. The land area and structure for the floor were prepared by John Wick, a local ranch owner who kindly provided the physical location, skills, and time to make sure that the floor would drain properly and the correct design was followed. This floor required digging out a ten-by-ten-foot spot of earth and filling it nine feet high with small, medium, and large aggregate rock. The final component of the floor was a four-inch layer of rice hulls, sand, and pounded clay.

Indigo is planted in late April to mid-May. The first harvest occurs in late June or early July when the crop is tall enough to prune. The second harvest occurs in late August, and then the plant is allowed to go to seed until mid-fall. After each harvest the leafed stems are laid out to dry, and with a watchful eye they are gathered at just the right moment when the stems remain supple and the leaves are crispy dry. To make sukumo, piles of the dried leaves and stems are stomped by human feet, the stems are removed and composted for nutrient cycling, and the leaves are saved until the end of the season. The pile is turned every seven days. The resulting composted leaf mass is then slowly fermented, a process that takes up to thirty days. When the vat is ready, the dyeing begins.

The requirements for producing sukumo's 440 pounds of dried leaf quickly transformed what had begun as a gardening effort into a small-scale farming project. It required approximately five thousand plants and about half an acre of land to grow the amount of indigo we needed for the compost process. It was the commitment to the color blue that has since inspired more than just one grower into this process; now there are multiple small-scale agriculture projects that have catapulted our community

Sally Fox's buffalo brown cotton fabric is worn underneath a cardigan that was handknit by Heidi Iverson (Honey Folk Clothing) and utilizes three strands of Fox's cotton, including one strand that is dyed in indigo.

into wider indigo production. This includes contributors from school gardens as well as small-acreage growers. Today multiple gardeners and farmers utilize the composting shed as a collective, pooling their dried leaf, turning the pile together, and divvying up the sukumo at the end of the hundred-day period. We are working on developing efficient planting and harvesting strategies to be able to inspire more growers to contribute plant material to this regionally focused indigo project.

I offer the indigo story as an illustration of a fibershed's strength in bringing community together. Over the years I've had the privilege of meeting and working with people from all walks of life and every aspect of textile making. I've taken part in learning journeys in Southeast Asia, the American Midwest, Navajo country, Scandinavia, Southern and Eastern Europe, and many other places to experience how other place-based fiber and dye communities function. I've had the honor of learning from diverse mentors, and in turn I've also had opportunities to share and teach what I've learned with a wide spectrum of collaborators. I've witnessed the growth of an international grassroots network that is changing the conversation about how and where we make our clothes, who makes them, and under what conditions they are made. My learning curve has been steep, and I am incredibly grateful for the helping hands, minds, and hearts that I have met along the way.

One of the profound lessons I have learned is that when we empower our communities to ground their livelihoods in material culture that is grown, processed, and utilized from their regional soils and in honor of the existing human cultural heritage, the strategies for how we design and implement solutions for many of our most pressing global challenges—including climate change and wealth inequality—become more precise and effective. These textile cultures by their very design are durable and resilient. We engage with and continue our ongoing learning from regional indigenous communities, whose understanding of the human role in the ecosystem is unparalleled. From these localized inquiries we've established what is an ever-growing understanding of the practices required to rebuild our soils, heal our climate, and strengthen our regional economies. The methods we've begun to utilize to generate globally impactful place-based solutions include developing bodies of site-specific research that help us measure the climate impact of soil-to-skin processes and develop appropriately scaled open-sourced technologies.

I have observed that there are mentors, willing students, and fiber and dye leaders at work in every textile community I've visited. There seems to be no lack of motivation among people to generate place-based textiles. What is needed to enhance these grassroots textile systems is the will and courage to appropriately resource and financially invest in them so that they can function to farm, ranch, mill, sew, repair, and then cycle materials into new clothes and eventually into compost. It's long overdue that we re-inspire our collective appreciation for clean and healthy textile systems and provide the appropriate levels of support to stabilize and lift this work.

About This Book

In the pages that follow, I will share the story and visual imagery of how we created a bioregional wardrobe from the plant, animal, and human skill within my home region and the effort it spawned to revive and regenerate textile traditions from fibers that haven't been grown for garments in over half a century. It is my hope that through an expression of place-based beauty, technical detail, and a reverence for the land, this story of how one fibershed has begun to define itself (and brief examples of others) will provide inspiration and a working tool kit for those who wish to replicate these efforts in their home communities.

At the same time the story is a call to action. We have a responsibility to evaluate new technologies that are being designed to offer "solutions" to our current textile systems. Humanity is at a clear inflection point where our fiber, dye, food, energy, and "you-name-it" systems have to be transformed if they are going to support us into the future. A brief look into textile history lays a foundation for understanding why technological advances that are adopted without precaution are not worth the risks. Perfluorinated compounds, for example, were heralded by industry for their stain resistance and water repellency, but they have now bioaccumulated into wildlife and human bodies worldwide and are known to have toxic effects to immune, liver, and endocrine system function. DDT was widely used on cotton until it was banned by the federal government in 1972. Designed to kill insects, the compound created a host of unintended consequences including, but not limited to, human autoimmune diseases. For example, girls exposed to DDT before puberty are five times more likely to develop breast cancer in midlife.[1] DDT is still present in our lakes, streams, and rivers and will persist for up to five centuries.

This book is designed to support the reader's critical thinking and evaluative processes. Together we need to ensure that the textile system transformations that we make do not shorten or debilitate our lives or the health of generations to come as well as the rest of the species on the planet. While the story is focused on fiber and natural dye systems, this book also taps into the broader cultural shift going on around the world as people and communities reconnect in a meaningful way with the land. With every passing day we increasingly see, hear, and feel the destructive effects of our complicity in perpetuating systems that were designed to make us the primary recipient of the planet's finite resources. *Fibershed* asks: How can we work together to transform contemporary cultural and economic systems to benefit all life and promote regeneration? And how can we do it without perpetuating consequences that force another set of technological solutions? As we learn the fundamentals of the carbon, water, and nutrient cycles, we understand that the earth's true ecological carrying capacity is directly connected to regenerative capacity of natural resources such as the soil and the fiber it grows. This knowledge begs a deep human question: How will we care for, protect, and use what the Earth provides in a manner that leaves the land and water more diverse and productive than we found them?

This book is designed to help you approach your own regional land base and inspire a commitment to work within its geography. It is an invitation to engage with all parts of the growing, creating, and wearing-and-caring processes. This can include reviving historical textile recipes that use long-forgotten fibers and dyes and creating new recipes that never existed. It is about making beautiful textiles from what we might perceive at first as limited resources. Hopefully the book will inspire new farming and ranching projects that will lead to increased and diverse offerings of new natural dyes and fibers and new income streams. If you're a textile entrepreneur developing your own brand or have spent years working at the corporate level of a major brand, you'll be able to use the latest understanding of biogeochemical processes and carbon cycling outlined in these pages to frame your fiber, dye, and supply-chain decisions. Implementing a deeper understanding of the earth's biogeochemical and physical properties and cycles is a critical step to generating new businesses that will help us remove the legacy load of carbon from our atmosphere (yes, clothing can help us do this!). The mainstream, industry-generated sustainability frameworks that have guided brands for years in their sourcing decisions have not been revised to incorporate the new science that has deepened our collective knowledge of the carbon cycle and the role of soil carbon sequestration. For that fundamental reason alone, there is a glaring need for companies to reevaluate their sourcing decisions with fresh eyes. Finally, this book is designed to help stimulate our work toward a regenerative future.

> **This book is designed to help you approach your own regional land base and inspire a commitment to work within its geography.**

Wool leg warmers from the 150-mile wardrobe,
knit by Allison Reilly with Mary Pettis Sarley's Twirl Yarn.

CHAPTER ONE

The Cost of Our Clothes

W hy do we wear what we wear? We often choose our clothes for comfort, for the way they look on our bodies, for their pleasing colors, for shelter from the weather, for protection from harm, for physical exertion, for entertainment, for time-honored traditions, for cultural expression, for group identity, and for the messages they send to our friends, colleagues, and co-workers. Clothing is so much more than just "fashion"—it sends important signals about our place in society, both actual and perceived, and is a critical element of our personal narrative—the daily, even hourly, choices we make to communicate our histories, desires, affiliations, and self-image. But when we think about what to wear, we don't usually consider the consequences of the manufacturing and disposal of our clothing. We also often overlook the ingredients in our clothes, including the type of fibers used to create a garment. Most of us don't even know what our clothes are made of, although the many toxic ingredients in them have a significant impact on our well-being. These ingredients (such as azo dyes and water-repellent chemistry) also have a major impact on the ecological status of the earth's natural systems of which we are part and parcel. And due to the sheer number of humans purchasing garments at the current rates of consumption, the cumulative impacts made by the ingredients in our clothing quickly add up. Over eighty billion individual garments were sold last year globally, a doubling in only fifteen years, supporting a $1.3 trillion textile industry that employs three hundred million people in nearly every country.[1]

What we wear is important, but until very recently there has been little public discussion of the significant environmental, social, and human health costs associated with the production, wearing, and disposal of our garments. The creation of textiles, including cotton farming, consumes nearly twenty-five trillion gallons of water annually, and 20 percent of freshwater pollution around the planet is attributed to the dyeing and treatment of garments.[2] The industry utilizes thousands of synthetic compounds, often in various combinations to soften, process, and dye our clothing, and many of which are linked to a range of human diseases, including chronic illnesses and cancer.[3]

The labor conditions in textile sweatshops and factories in most cases offer income that provides little opportunity for the kind of social mobility many people in developed Western nations have come to expect.[4] In 2015 six thousand garment workers in Cambodia protested for fair wages and improved working conditions. They were joined by workers in India and other countries—all part of the supply chain that produced clothing for H&M, one of the biggest clothing retailers in the world.[5] A 2013 report by the International Labour Office stated that 168 million children worldwide are considered to be child laborers, almost 11 percent of the population as a whole. Though most worked in the agricultural sector, many labored in the textile industry.[6] In India nearly half a million children work in the nation's extensive cotton fields.[7]

These poor labor conditions are exacerbated by our clothing consumption. According to Elizabeth Cline, author of *Overdressed: The Shockingly High Cost of Cheap Fashion*, in 1990 half of all garments worn by Americans were made in the United States. Today that figure stands at only 2 percent.[8] According to a Greenpeace analysis published in 2015, the average person today is buying 60 percent more garments than he or she did in 2000 and keeping them for half as long. Global demand for clothing, particularly in Asia and Africa, is projected to double by 2050. The turnaround time for fashion trends—the rate that we use and throw away clothes—dropped by 50 percent between 1992 and 2002. Some turnaround times today are as short as two weeks, a trend that has been dubbed Fast Fashion. "We buy more clothes than ever before [and] we wear them fewer times," the Greenpeace authors wrote. "By treating clothes as disposable items, fashion has become a novelty, and the commercialisation and marketing of fashion is leading to overconsumption and materialism—keeping our clothes and cherishing them is not in fashion anymore."[9]

> These poor labor conditions are exacerbated by our clothing consumption.

Two brands closely identified with Fast Fashion, Zara and H&M, together produce one billion items per year, a large portion of which are thrown away after only a few wearings. Only 15 percent of used clothing in the US is recycled; the rest winds up in landfills, more than 5 percent of all the municipal waste generated annually.[10] By 2019 the total amount of textiles purchased is expected to exceed thirty-five billion pounds. Another study pegs the value of this discarded material at $460 billion per year.[11] The authors of the Greenpeace report estimate that as much as 95 percent of all clothes thrown out could be reworn, reused, or recycled.[12]

The fibers in our clothing also have alarming narratives. Polyester, an oil-based synthetic fiber, is used in 60 percent of our garments today, more than double the amount used in 2000. It consumes nearly 350 million barrels of oil every year and accounts for 282 billion kilograms of carbon dioxide emissions, three times higher than the amount for cotton, which will make it increasingly difficult for the world to meet the two-degree Celsius climate goals set down in the Paris Agreement. Non-oil-based fibers come with a cost as well. Rayon, which doubled in use between 2005 and 2015, is a cellulosic fiber that requires the harvesting and processing of trees and bamboo. Canopy, a Canadian nonprofit organization, is working to ensure that threatened forests are not part of global rayon supply chains, and significant gains have been made to protect endangered forests. However, the primary issue remains—using tree pulp for clothing is a land- and fossil-fuel-intensive process that puts our precious global forests at risk. Then there is the conventional processing method whereby tree and bamboo pulp is dissolved in carbon disulfide, exposing workers to poisonous fumes that can cause severe neurological damage. The process of manufacturing rayon is so toxic that it was banned in the United States in 2013 by the EPA. A report in 2006 by the European Commission's Joint Research Centre designated acrylic, a synthetic fiber, as one of the most toxic fabrics to produce in the world.[13]

In addition, microfibers from our synthetic clothing have been shedding into our waterways for decades, making their way into our oceans and onto our soils. In research sponsored by clothing and gear company Patagonia, a team of graduate student researchers from the University of California–Santa Barbara's Bren School of Environmental Science and Management investigated the scope and impact of this issue. "Our research found that microfibers are prevalent in both aquatic and terrestrial habitats, from the bottom of the Indian Ocean to farmland in the United States," the team wrote on its website. Their experiments found that when synthetic jackets are washed in a machine, approximately 1.7 grams of microfibers are released and travel to the local wastewater treatment plant, where up to 40 percent of them enter into rivers, lakes, and oceans.[14]

A separate study, led by researchers from the University of California–Davis, examined the presence of microfiber pollution in ocean fish, discovering that one in three shellfish, one in four finfish, and 67 percent of all species tested from fish markets in California contained micro-fibers.[15] It is further estimated that Europeans ingest up to eleven thousand pieces of microfiber per year through shellfish consumption.[16] The Bren School research team concluded that retrofitting wastewater treatment plants to capture microfibers was too expensive and should not be considered the strategy of choice to mitigate the pollution. The team also noted that once marine organisms ingest the microfibers, they some-times end up starving and may also suffer reproductive consequences, partly because microfibers attract other toxins, including DDT and PCBs.

A Brief History of Clothing

It is important to note that for most of human history, it wasn't this way. Once upon a time all our clothes were "organic." When our ape-descended ancestors lost their body hair one million years ago, they apparently went about their daily lives naked until the emergence of *Homo sapiens* roughly two hundred thousand years ago. In 2011 a genetic study of the evolution of human body lice, which require the presence of clothing as shelter to live, pegged the first use of clothes at 170,000 years ago.[17] This first wardrobe likely consisted of animal furs and leather draped or tied around various parts of our bodies. Soon humans—as well as their Neanderthal cousins—began poking holes into their clothing with sharp bones called *awls*, which enabled them to bind garments tightly together with leather cords in order to keep warm and create longer-lasting clothes. This was handy for the human populations who dispersed out of Africa sixty thousand years ago, allowing them to colonize much colder environments. Since *H. sapiens* was leaner than the stocky Neanderthals, thus needing more protection during the ensu-ing ice age, the creative use of clothing-making tools may have given a competitive advantage.

Around forty-five thousand years ago, an enterprising individual or band, likely living in Central Asia, had the bright idea to create a small hole in the end of a bone awl, thus inventing the *needle*, one of the most underrated technological developments in human history. The first nee-dles employed a thin thread composed of animal sinew or plant fiber, which allowed articles of clothing to be constructed, repaired, or repur-posed in new ways. In prehistoric France there is evidence that the cave-art-making Cro-Magnon people made close-fitting hoods, shawls, pants, and shirts from diverse animal hides sewn together. This ward-robe continued mostly unchanged for forty thousand years, confirmed by the discovery in 1991 in the Austrian Alps of Otzi the Iceman, a five-thousand-year-old mummified man. His well-preserved clothing, including finely sewn leggings, sophisticated shoes (with leather shoelaces), lengthy coat, and soft loincloth, were made from many different types of animal hide and fur. He also carried a leather pouch attached to a belt that contained, among other useful items, a bone awl, possibly for repairing his outfit as he traveled.

The Agricultural Revolution, which began approximately ten thousand years ago (and the subsequent domestication of fiber-producing livestock), changed our clothing dramatically. Thread spun from farmed wool and linen became available, and as it became increasingly refined, it enabled the creation of *cloth*. The introduction of bronze needles around five thousand years ago meant sewn clothing became much more sophisticated and widespread. However, needle technology wasn't advanced enough for fine sewing. The amount of time and labor required to produce a single piece of cloth to cover a human body was so large that early civilizations in India, Greece, and Egypt preferred one-piece garments, such as togas, saris, and tunics. Cloth pants were apparently invented by horse-riding cultures in Central Asia, who also created quilted jackets. Even the development of iron sewing needles and new spinning technology during the height of the Roman Empire didn't diminish the desire for one-piece clothing. Dyes at this time, including the famous royal purple, were derived from plants and insects, some exotic, and their acquisition stimulated commercial trading enterprises as well as imperial ambitions.

Clothing over the subsequent centuries diversified significantly as cultures developed their own styles, fabrics, and dyes. Its footprint on human and ecological health, however, remained relatively small. While prehistoric agriculture had a variety of damaging environmental consequences in many parts of the world, and slavery was often employed by landowners and other elites to produce clothes, the raw materials themselves remained organic (in the modern "non-synthetic" sense).

The Industrial Revolution changed everything. Textile manufacturing was one of the very first industries to be mechanized, and the social, economic, and working conditions involved in making clothes underwent profound and often disturbing changes, setting the industry on the road to modern textile production. During the discovery of oil and the subsequent advancements in synthetic chemistry in the second half of the nineteenth century, the groundwork was laid for the negative impacts that our clothing and textile production systems are still having on our personal and planetary health. It is a price that we continue to pay to this day.

The Dangers of Synthetic Chemistry

For nearly the entire time *H. sapiens* has existed on Earth, the chemical makeup and concentrations of organic compounds within our biosphere have been generated from a complex set of variables. You could consider it a multibillion-year conversation among planetary natural forces that resulted in our complex and diverse planet. Today humans continue to ingest, inhale, and physically interact with their material surroundings, but the chemical composition and concentration of the compounds being absorbed by our bodies have greatly changed. We now live with a host of new synthetic chemicals with which we did not co-evolve, and we have become exposed to artificially high concentrations of toxic compounds. But the makers and producers of these compounds have developed, marketed, and promoted their widespread use without engaging third-party precautionary testing to determine their potential human health and environmental impacts. In fact, of the nearly seventy thousand synthetic compounds in use today, less than 2 percent have been tested for their impact on human health.[18]

There is no federal law that mandates third-party safety testing for approval of the chemicals that are used in our textiles, cosmetics, toys, art supplies, carpets, or construction materials. The Toxic Substances Control Act, passed in the United States in 1976 and updated in 2016, now requires that the industry supply safety data to the EPA, and in turn the agency will use that data to determine if a chemical is a high or low priority. High-priority chemicals undergo a risk evaluation, paid for and overseen by the industry unless the EPA is specifically requested to take part in the research.[19]

Due to the ease with which these compounds have historically entered our economy and thus our lives, it is not surprising that the vast majority of we humans now carry these compounds in our blood, urine, fatty tissue, amniotic fluid, bone, sperm, and breast milk. Based on the way our laws are designed, it is only in hindsight that our society has had the wherewithal to identify and remove the compounds that cause

Several industries, including dye factories, are located upstream from this site on the Tullahan River in the Philippines. *Photo by Gigie Cruz-Sy / Greenpeace.*

serious harm to the living world. Author and biologist Sandra Steingraber discussed the concept of toxic trespass in a 2013 interview with Bill Moyers. "We inhale a pint of air with every breath," said Steingraber, and in those inhalations, "we take in chemicals without our consent."[20]

More is involved than just the air we breathe. Our skin, the largest permeable organ in the human body, absorbs material surroundings and deposits toxicity directly into our bloodstream. But it takes herculean legal action and often decades to properly address the toxic trespass of even *one* chemical compound. Even when a chemical compound is proven by multiple studies to take a costly toll on our ecology, we must gather lawyers and scientists; only if every facet comes together are we *sometimes* provided an opportunity to say: "Thank goodness we've stopped using DDT in our neighborhoods and farms," or "Good thing we took lead out of our fuel and paint," or "What were we thinking when we thought cigarette smoking was good for our health?" It's very likely that we will soon be saying, "Why did we wear conventionally farmed or fossil-carbon-derived clothing?" And "How could we have ever gone without an ingredients list on our garments?" We have blindly been wearing carcinogens, neurotoxins, and endocrine disruptors on the biggest organ of our body—*how crazy is that?*

Geana Sieburger's studio,
GDS Cloth Goods.

The future is rushing at us, but there is still time for significant work to be done to achieve transparency about what we wear and help spread knowledge and information. And while transparency and public education are powerful tools for cultural transformation, it is also important to ask larger questions about the efficacy of a legal framework that allows so much damage to be done upfront, with the burden of proof left to a combination of the chronically sick and dying, and a community of altruistic experts willing to work with very little pay to fight uphill legal and scientific battles. Why do we, as a cultural and economic system, allow this toxic trespass in the first place? This is a fundamental societal question.

It is the belief of many doctors, researchers, and scientists that the general public needs to become legally protected from chemical trespass that continues to occur without our consent. It would not be difficult as a society to identify hazardous synthetic compounds and eliminate their production, use, and release. This change could occur swiftly and in less than one generation by developing laws that require third-party precautionary health impact analysis on all new chemical compounds entering the marketplace and through full disclosure on the impacts of any existing and persistent synthetic chemical compounds that will likely remain in our environment through lingering remnants of material culture (shelter, durable goods, garments).

The larger transformative goal put forth by Sandra Steingraber is to develop an economy that no longer depends upon these compounds, requiring us to focus collectively on profitable and ecologically beneficial alternatives. In the meantime we will continue to live in a biosphere where nearly one-quarter of all deaths across the globe are directly caused by living and working in toxic and polluted environments, and more than one-quarter of the deaths of children under five are attributed to environmental conditions.[21]

In the last decade researchers have observed a significant shift away from infectious, parasitic, and nutritional diseases to *non-communicable* diseases, meaning non-infectious. This shift is most noticeable in non-industrialized nations where communicable diseases were historically a common cause of death. Non-communicable diseases, such as those generated by environmental toxicity, have now been identified as "one of the major challenges for sustainable development in the twenty-first century."[22] The World Health Organization stated in their report *Preventing Disease Through Healthy Environments*: "The most effective and sustainable measures to combat the environmental causes for disease need to be designed and implemented holistically."[23] The report emphasized the value of creating healthy environments and stated that

Notes from a Textile Designer

by Georgene Shelton, textile designer and
supply-chain consultant who works with Fibershed

In the course of working with printed textiles development, I was exposed to a lot of unfinished textiles, and I handled a lot of what are known as strike-offs (first-pass textiles). I developed a mysterious rash during this work that was with me for a few years, which started on my wrists, hands, and ankles. Dermatologists in the United States could not figure out what it was and continued to call it contact dermatitis. Eventually I went to a dermatologist in New Delhi who deemed it to be "Lichen Planus." He gave me some of his handmade herbal pills and recommended that I tell my employers to give me some time off so I could relax. Thus I was able to take a few weekends to travel outside New Delhi. I went to Jaipur, Jodhpur, and Udaipur and was able to see some of the weaving, dyeing, and printing at a much more primitive local level than I usually got to see.

I was also exposed to a considerable amount of textile outgassing of a particular chemical in rayon manufacturing. I have to wonder how much I was exposed to, as I was knee-deep in the stuff. My travels in India were done for work: This included the hotel, the factory, and the road in between. I was in areas that tourists never see, where the ditches run green with industrial runoff, and hairy curly-tailed pigs cavort in garbage-strewn vacant lots with grayed fluffs of textile waste caught in the rusty barbed-wire fences. Much of the time I traveled in moto-rickshaws, those little two- or three-person black-and-yellow taxis that are mounted to motorcycles. Being stuck in traffic, with the thick fog of exhaust pollution, was unbearable. Add to that the smokestack pollution of a city of textile mills and it's a wonder that I, or any of us, survived. We're still here, but not without health trade-offs.

this effort should become a primary focus for disease prevention in the twenty-first century. Protecting the environment is a means to protect human health.

Coordinating and taking action across sectors (energy, industry/manufacturing, water and sanitation, agriculture, housing, transport) will be necessary if we are to address the environmental causes of chronic illness. Acting together on coordinated health, environment, and development policies can strengthen and sustain improvements to human well-being and quality of life through multiple social and economic co-benefits, repositioning the health sector to work on effective preventive policies. This is the way forward to address environmental causes of disease and injury and ultimately to transform the global burden of disease that has reached epidemic proportions. The rate of childhood cancer rose 10 percent between 1973 and 1991; the rate of testicular cancer has doubled in the last twenty years; the likelihood of a woman developing breast cancer in her lifetime jumped from one in twenty in 1960 to one in eight in 1999.[24] The textile industry, from farm to skin, relies on a host of substances that are now known to cause reproductive disorders, cancer, infertility, DNA damage, behavior problems, and endocrine disruption, to name a few of the most notable issues (see appendix A, page 227, for more information).

A majority of Americans have expressed concerns regarding their exposure to health-harming chemicals. In a recent poll taken to accumulate data on voters' viewpoints on the Toxic Substances Control Act, 87 percent of the respondents said they were concerned that chemicals had entered the marketplace without proper testing, and 66 percent said they were *very concerned*.[25] "What's really surprising is that voters across almost all demographic groups—Democrats, Republicans, and Independents—said that they didn't think that regulations on chemicals are strong enough," said pollster Celinda Lake. With this much bipartisan support, the policy pathway to protecting our environment and our health (which are one and the same) is strong enough to move our society toward instituting long-overdue third-party research and safety testing.

It is important to note again that we produce only *2 percent* of the American wardrobe from within our own political borders. While we have banned many of the most egregious compounds from being produced in our nation and home communities, we continue to import the residues of toxic compounds used in global textile manufacturing processes directly into our closets, and soon after into our skin and lungs. Our consumption of these toxic goods continues to promote unhealthy exposure levels to global textile-producing communities—putting many lives unnecessarily at risk. We have to ask ourselves: If we would not want to work in fiber and dye processing facilities that disproportionately expose us to chemical compounds that cause premature deaths, debilitating diseases, and DNA damage that impacts our children's and grandchildren's health, where is the justice and equity in asking others to do this on our behalf? The conscious and unconscious levels of entitlement that our opaque and off-shored textile system has created are unconscionable.

Synthetic Dyes and the Cost of Color

More than eight thousand different synthetic dyes are used in the production of our garments, many derived from fossil carbon sources.[26] Historically these compounds are based upon the scientific innovations of William Perkins, a chemist who discovered the first synthetic aniline

An oil derrick (where synthetic color is derived). *Photo by Joshua Doubek.*

dye during an attempt in 1856 to create a cure for malaria. Synthetic dyes are a relic of the Industrial Revolution; originally it required four hundred pounds of coal tar for a four-ounce yield of blue dye. Today's synthetic dyes are responsible for the color of textiles all over the world (minus a small percentage of undyed and naturally dyed garments). The most ubiquitous colorants, responsible for 60 to 70 percent of the total of industrial practices, are azo dyes, a number of which have been demonstrated to be both carcinogenic and mutagenic (meaning they disrupt the DNA in our cells).[27] There are currently twenty-two azo dyes that the European Union does not allow for use on textiles that are worn directly next to the skin.[28]

Two hundred thousand tons of dye are left unbound to textiles globally and are lost to effluent.[29] When allowed to enter freshwater aquatic systems, these coloring agents cause oxygen deficiencies and can heavily impact drinking and irrigation water.[30] In a 2010 study three synthetic dyes were identified in rivers in various concentrations, and the resulting drinking water sourced from the river contained these dyes even after treatment. This showed that methods such as flocculation, coagulation, and pre-chlorination were not enough to clean the water for human

A garden of coreopsis, a favored dye flower for creating orange and yellow dye colors.

consumption.[31] Simply put: Dyes do not break down easily, and they resist biodegradation, which is why they persist in our waterways.

In an interview with Dr. Shanna Swan, professor of environmental medicine and public health, obstetrics, gynecology, and reproductive science for the Icahn School of Medicine at Mount Sinai, the subject of aniline dyes came up. "We have far more data on compounds that are ingested, such as pharmaceuticals," said Swan, "because we take more responsibility for testing their impact on human health. We have far less data on substances that are not designed to be ingested, but what we have realized via extensive research is the impact of the painkiller acetaminophen/paracetamol (branded as Tylenol and other name brands) on the human endocrine system."[32] Swan went on to say that following exposure to aniline (a component of textile dye), the body metabolizes it into acetaminophen (N-acetyl-4-aminophenol or NA4AP), which is known to pass through the placenta. It has also been detected in human breast milk and is a known endocrine disruptor. In a German study aniline was present in over *90 percent* of the population.[33] What we still do not understand is the rate and impact of aniline being absorbed by the skin or inhaled via lint—but these are areas for more research.

In one human monitoring study, the subjects were provided a fast from oral exposure to both aniline and acetaminophen—and yet the continued and ubiquitous excretion of NA4AP (the metabolite for aniline) was noted in urine samples.[34] The same study summarized the issue by saying that the consistent and ubiquitous presence of aniline within the general population warrants further investigation. The question remains: If we are not ingesting aniline orally, yet urine samples continue to show evidence of exposure, how are we being exposed? NA4AP has been shown to cause health problems at the most sensitive moments of human life. When a fetus is exposed to anti-androgenic endocrine disrupters, we see issues such as reduced anogenital distance, smaller or disfigured penises, and behavior shifts that show up as "less male" play behavior in young children.[35] "Eight weeks is a critical time for a fetus to differentiate itself as a male—testosterone is required for this transformation, and if there are any impediments or changes in the levels of this hormone due to a mother's exposure to these anti-androgenic compounds—it can create the kinds of physical changes and mutations we are seeing in far greater number now," said Dr. Swan.[36]

We Are Wearing Endocrine Disruptors

The most common suite of compounds of concern that show up in the textile supply chain are endocrine disruptors (EDCs). Estrogenic and anti-androgenic compounds have been found in the fields of genetically engineered cotton, washing agents, dyes, and textile finishes and coatings.

> Estrogenic and anti-androgenic compounds have been found in the fields of genetically engineered cotton, washing agents, dyes, and textile finishes and coatings.

Why should we care that synthetic compounds are changing the production, transport, storage, and synthesis of our hormones? If you consider that the endocrine system produces complex chemical compounds (hormones) that regulate our metabolism, sleep, mood, growth, development, reproductive health, and tissue health, the answer is simple. The endocrine system is responsible for the foundational organizing principles of life and is often described as the communication system that functions both inside and between cells.

At the start of the twenty-first century, we began to refine our understanding of the impacts of EDCs on human and wildlife health. Scientific reviews published by the Endocrine Society, the European Commission, and the European Environment Agency provide a scientific basis for concern over the impact of EDCs on both human and wildlife populations.[37] Two agencies focused solely on the health of our youth, the European Society for Pediatric Endocrinology and the Pediatric Endocrine Society, have put forward a consensus statement calling for action regarding the regulation of endocrine disruptors due to their health impacts on developing fetuses and young children at what are known to be critical stages of sexual development.

One of the primary concerns of medical experts is the rate at which incidence of disease has risen in the last several decades, a rate that rules out genetic factors as the only plausible explanation according to a 2012 report by the World Bank and the United Nations Environment Program titled *State of the Science of Endocrine Disrupting Chemicals*.[38] A study on reproductive health with a sample population of 26,600 men found that sperm concentrations had decreased by a third since the 1990s—that's a 32 percent decrease in sperm concentrations over a seventeen-year time span.[39] The incidence of genital malformations (non-descending testes and hypospadias) has increased, and the global rate of endocrine-related cancers, including endometrial, ovarian, breast, prostate, testicular, and thyroid, has increased in the last forty years, as have the rates of obesity and type 2 diabetes.[40] There is a long list of studies that tie EDCs to these diseases, as well as establishing their detrimental impact on male and female thyroid disorders, neurodevelopmental disorders in children, metabolic and adrenal disorders, and bone and immune function diseases.[41] The UN and World Bank report is also explicit about EDCs' connection to wildlife health. "Given our understanding of EDCs and their effects on the reproductive system," the authors wrote, "it is likely that declines in the numbers of wildlife populations have occurred because of the effects of these chemicals on these species."[42]

The report also discussed endocrine disrupting impacts on the quality of genetic information that is passed down between generations. This area of study is known as *epigenetics* and is defined as the "heritable changes in the genome that are not dependent upon changes in genetic sequences."[43] According to the study, "A number of these chemicals have now been shown to influence epigenetic mechanisms and to produce effects in several generations of animals." It is plausible that these epigenetic impacts are occurring within humans as well. There is mounting evidence and concern that human exposure to EDCs in a parent generation is impacting the health of second and third generations. The question for us as clothing wearers is this: Is it worth creating

Transparency in Ingredients

There is nothing quite like transparency to bring about attention, conversation, and decisive action. We must ask ourselves: Why don't we have a chemical-specific ingredients list on our clothing? Armed with real knowledge, wearers could begin to carry out consumer decisions in a proactive manner, just as they do when buying groceries. For many of us, the current textile certifications remain mysterious. There remains a general lack of understanding about what clothing certifications mean and how they physically relate to personal health and ecosystem function. Activists hope to initiate a conversation around the health impacts of these compounds. Starting from a place of knowledge, the public can begin to collectively strengthen their understanding and thus their relationship with the supply chains that produce our textiles. As educated wearers, we can drive the creation and proliferation of nontoxic clothing by focusing our purchases on solely those items we *know* are safe. The more access to knowledge we have, the more empowered our actions become; if we work collectively we are assured positive outcomes to our health and the health of the ecosystems we depend upon.

immediate and/or long-term intergenerational DNA damage to ourselves and the species we share this planet with so that we can be assured of filling our closets with brightly colored, stain-free, and wrinkle- and water-resistant clothing? If the answer is no, then we must ask ourselves what we can do as wearers to protect both our personal and our collective planetary health.

Fast Fashion and Other Troubles

Not long ago the fashion industry had two seasons: spring/summer and fall/winter (which actually took place in September and February, respectively). But as globalization took hold in the 1990s and online shopping on the internet exploded in the 2000s, the industry saw dramatically shortened manufacturing supply lines and "shelf" life of fashion styles. Their response was to add more seasons to the fashion calendar and accelerate the introduction of new lines. This trend was abetted by a surge in global tourism, which created year-round demand for all kinds of clothes for warm and cold weather, further blurring traditional boundaries between seasons. Then, starting around 2000, came Fast Fashion—the popular name for clothing that is deliberately designed and marketed to go out of style after only a few days or weeks. Fast Fashion's hallmark was disposable, low-quality garments derived from fossil-fuel-sourced synthetic fibers mass-produced in poorly regulated factories under hazardous working conditions, generating copious amounts of toxic waste and shipped long distances to stores to be sold cheaply and then discarded after a few weeks. As a result of the onslaught of cheap, disposable clothing, the fashion industry now boasted as many as fifty-two microseasons per year, cranking out new styles endlessly, trying as hard as possible to get customers to buy as many garments as quickly as possible.

It worked. Between 2000 and 2014 global clothing production doubled. The average person buys 60 percent more garments than they did fifteen years ago and keeps them half as long. On average, an American buys sixty-seven items of clothing every year, which is roughly one item every five days.[44] Fast Fashion was pioneered by Zara, a mammoth designer and retailer based in Spain, which began to make deliveries from the factory to its stores once, then twice, a week. Today some retailers receive shipments of new styles *daily*—and that doesn't even include the rapid turnover of clothing made available online. Meanwhile, the amount of clothing that Americans throw away each year has doubled over a decade to almost eighty pounds per person.[45] Almost none of it is recycled or resold; most of it winds up in a landfill. Most people don't think of what happens next: Garments made from oil-based synthetic fibers can take up to two hundred years to decompose. And sadly, donating clothes to Goodwill or another charity won't help much. Many "resale" garments are exported to developing nations, where only 30 percent are resold due to poor quality. The remainder are reprocessed into cheap yarn or tossed out as waste. Recently a number of nations have even begun to impose restrictions on imported used clothing, including outright bans.

> Garments made from oil-based synthetic fibers can take up to two hundred years to decompose.

The Fast Fashion business model, which depends on thin margins and high volume, has become profitable by offering trendy items to shoppers on a consistently changing basis—in other words, providing new and novel products with rapid inventory turnover. Retailers have created, among many consumers, a desire for constant new wardrobe inputs, a behavior that is maintained by throwing garments aside at equally fast rates. "A store like H&M produces hundreds of millions of garments per year," author Elizabeth Cline said in an interview on National Public Radio in 2013. "They put a small markup on the clothes and earn their profit out of selling an ocean of clothing."[46]

Thanks to smartphones, many bulk manufacturers such as Zara and Primark get their knockoff designs from images captured on phones by attendees of mainstream fashion shows (frequently generating charges of intellectual property theft). This feeds the desire for instant gratification that has come to define much of the global economy in the internet era. Fast Fashion is made possible by a system of "innovations" in supply-chain management intended to keep costs as low as possible no matter what—a system that often exploits workers, the environment, and neighboring communities. The faster Fast Fashion becomes, the greater the negative impacts it has locally and globally. And these impacts are set to grow dramatically. By 2050, based on current trajectories, the textile industry will use as much as 26 percent of the global carbon budget associated with the two-degree Celsius pathway set out in the Paris Agreement. The amount of non-renewable resources consumed by the industry will triple over the same time span.[47]

Embedded in the Fast Fashion production model is the exploitation of human labor. Poor factory conditions, long hours, and low wages have long been a staple of the industry, but in recent years Fast Fashion retailers have been accused of knowingly buying garments made by prison labor and children, firing pregnant employees, and suppressing workers' rights. This situation is not unique. It is estimated that one-sixth of the world's population is employed making textiles, but since over 98 percent of America's closets are filled with clothing made in *other* countries, many of us are no longer exposed to the realities of industrial textile production and processes. We never see the direct impacts the industry has on those who make our clothing. The reality is that men, women, and children who work in the industry are exposed to much higher doses of carcinogenic, neurotoxic, and endocrine and immune-system disrupting chemicals, and higher rates of colorectal, thyroid, testicular, bladder, and nasal cancer have been documented in textile workers.[48] There are many points within the supply chain where exposures to health-harming compounds can occur, whether it's handling textiles in the agricultural and oil fields (in the case of natural or synthetic fibers) or handling the compounds that pass through human-run systems during washing, combing, spinning, knitting, weaving, and finishing. Those who repeatedly handle bolts of textiles fabrics, including designers, have also been known to contract adverse health conditions from freshly manufactured cloth.

Working conditions can be dangerous in other ways. On April 23, 2013, Rana Plaza, an eight-story building that housed five separate garment manufacturers located outside Dhaka, Bangladesh, collapsed, killing eleven hundred workers and seriously injuring two thousand

Children working in a ginning factory in India, where cotton is separated from seed. *Photo by Joerg Boethling / Alamy Stock Photo.*

more. Major cracks in the structure had appeared the day before the tragedy, but the factory owners browbeat their employees to continue working even though a government inspector had ordered the building to be evacuated. A few months earlier a fire at another factory claimed the lives of over one hundred garment workers, one in a series of fires in Bangladeshi textile factories. In both cases (plus hundreds more) factory managers had denied workers the right to safe working conditions, sometimes even placing physical barriers at exits from perilous structures to deter employees from fleeing.

A 2015 report by Human Rights Watch detailed a pattern of violations, including illegal anti-union activities, assaults on organizers, physical abuse, sexual intimidation, forced overtime, denial of maternity leave, failure to pay wages and bonuses, contaminated drinking water, and loss of employment often at the whim of a supervisor. "Almost all of the factories make garments for well-known retail companies in North America, Europe, and Australia," the author wrote. "Violations of workers' rights were a problem in nearly all of the factories and included practices contrary to both national Bangladeshi law and codes of conduct that Western retailers insist, often in production contracts, that their suppliers follow."[49] News of these violations, however, does not often make it to the consumers of the products made in Bangladeshi factories unless there is a headline-grabbing tragedy. The irony is painful—citizens of Western democracies are ignorantly buying garments produced by workers in repressive conditions that border on slavery (this story isn't unique, as the history of cotton demonstrates; see chapter 5).

China, which produces more than 50 percent of the world's textiles, is another example of the cost of our clothing. China produces more greenhouse gases than any other nation. Many of its major cities suffer air pollution levels that far exceed the threshold for human safety. More than one hundred Chinese cities grapple with persistent water shortages, and nearly one-third of the rivers have been designated as too polluted for human contact. Meanwhile, textile manufacturing consumes as much as eight thousand gallons of water for every eleven thousand yards of fabric produced, or more than ninety-five billion gallons of water each year, representing the third-highest rate of water use among Chinese industries after chemicals and paper. It is also among the most energy-intensive industries in the nation, largely powered by coal. China is the largest consumer of textile chemicals in the world, many of which are discharged directly into rivers and streams (often at night to avoid detection), contaminating drinking water supplies and creating a toxic environment for fish and other forms of aquatic life. An investigation by Greenpeace at two major Chinese clothing manufacturing sites discovered a wide range of hazardous chemicals in the river below the factories, including alkylphenols, which are banned in Europe, and perfluorinated chemicals (PFCs), in particular perfluorooctanoic acid and perfluorooctane sulfonate.[50] Greenpeace traced the chain of custody for the clothing back to the global brands that bought the garments from these factories and confronted them with the evidence. Greenpeace investigators examined clothing from other nations as well, finding similar challenges.[51]

> The first step in resisting this trend is to wear our clothes longer, even to the point of wearing them out completely.

The Slow Clothing Solution

The primary way to reverse this seemingly irreversible reality begins with us, the wearers. Quite simply our consumption needs to slow down. Fast Fashion is new to our civilization, and there is no reason to retain it. The first step in resisting this trend is to wear our clothes longer, even to the point of wearing them out completely. By doing so, we can learn to care for those items that need mending, which involves wonderful cultural practices. Cycling clothing through our communities by shopping at consignment shops, Goodwill, Salvation Army, vintage boutiques, and clothing swaps is also a method for keeping clothing in play, as long as it doesn't mean buying more "fast" clothes and perpetuating the cycle. At the corporate brand level, we've seen Eileen Fisher's take-back program, Green Eileen, blossom into a Brooklyn pop-up shop with the theme "Remade in the USA," where the company produced a limited five-hundred-piece collection in collaboration with their Social Innovator Award winners. Patagonia's Worn Wear program provides customers the option of bringing their garments back to the store to be repaired and resold through the Worn Wear website, and if the clothing item is in good enough condition, the customer will even receive store credit. The North Face's Close the Loop program ensures that our clothing won't end up in a landfill, and offers drop-off points in their stores for all garments and footwear; the company will accept all items (from all brands).

A wonderful example of a culturally transformational experiment that re-envisioned the modern wardrobe is Katrina Rodabaugh's Make Thrift Mend project. She described her work:

> *On August 1, 2013, I launched my fast-fashion fast, Make Thrift Mend, and vowed to abstain from factory clothing for one year while I focused on making my own garments, buying second-hand, and mending what I already owned. Conceived as an art project examining the intersection of sustainability, fashion, fiber art, and what's known as social practice or "art as action," Make Thrift Mend was initially a personal art project. In that first year I made simple garments, became better educated about fiber contents and origins (and quickly prioritized biodegradable secondhand clothing like linen, cotton, silk, and wool, which also takes natural plant dyes beautifully), scoured bookstores and websites for all the information I could find regarding slow fashion, and taught myself to mend, forage for plant dyes, and better care for or rejuvenate existing clothing. I hosted mending parties, received a grant to teach a free online class, led a mending workshop, and continued to educate myself on sustainable fashion practices.*[52]

Rodabaugh now teaches mending, natural dye process, and slow textile ethos as a way to rejuvenate cloth and garments.

One of the leaders of the slow fashion movement is Kate Fletcher, a professor of sustainability, design, and fashion at the University of Arts

Katrina Rodabaugh's mended garments.
Photo by Katrina Rodabaugh.

in London, England. A pioneer of the "craft of use" concept, she advocates for an innovative approach toward wearing our clothes that is based on mending, reusing, and repurposing in imaginative and fashionable new ways. *Use* in her sense means so much more than simply buying a piece of clothing, wearing it, and then disposing of it when you're done. While there is much expertise that goes into the crafting of a garment, rarely is equal skill brought to bear on its use, particularly the necessity of extending the garment's life. Fletcher calls this approach *post-growth*, because it means thinking about your garment in the context of sustainability, added value, consciousness raising, and learning new skills, such as mending with needle and thread. She has widely published her thoughts and designs and recently gathered together many examples of ingenious and practical use of garments from around the world into a book titled *Craft of Use*. Her goal is to encourage readers to "adopt a more ecological idea of fashion that recognizes what happens beyond design and production as rich, powerful, [and] valuable."[53]

Part of caring for our clothes means washing them. And while we all like to be clean and smell good—or at least not smell offensively—we should wash our clothing less; when we *do* wash it, it's best to use cold water and mild soaps. This keeps our clothes in play longer and saves freshwater resources (also, using cold water saves us from having to heat water, which in most cases is done with propane or electricity). For our own personal and our families' endocrine health, it is vital to stay away from synthetic fragrances in our detergents, which are 95 percent petrochemical-derived and contain endocrine disrupting phthalates. These plasticizers keep fragrances on our clothes long after we've washed them. If you've ever been hiking or trail running in the purity of open air and you encounter someone whose clothing emits the odor of laundry detergent, you are smelling a fragrance held to the garment via an endocrine disrupter. The fragrances themselves can be made up of over six hundred synthetic chemicals; it is a challenge to uncover their molecular makeup, not to mention that the surfactants and water softeners found in most of our laundry soaps are known skin irritants and or cell disrupting agents. Look for non-fragrance detergents that are designed to break down in gray-water systems (household water systems that move laundry water into our gardens). There are very good recipes for DIY laundry detergents that are more affordable than any store-bought soap on the market.

Once our clothes come out of the washer, line-drying them is the lowest-footprint method for evaporating out the water. When you're caring for wool outerwear, keep in mind that wool sweaters, coats, and any outer-layer garments can be cleaned by exposure to ultraviolet light.

Composting Our Clothes

Natural fiber garments come from the soil and therefore can return to the soil. Hemp, cotton, wool, alpaca, cashmere, nettle, ramie, silk, mohair, angora, qiviut, guanaco, yak, and flax are all examples of natural fibers that can be transformed into compost when their life cycle as clothing is over. Like food scraps and yard clippings, natural fiber clothing is composed of proteins (animal fibers) or carbohydrates (plant fibers) or a blend of the two. The carbon-based molecules that are in our food are also in our fiber. If our clothing items are 100 percent natural and not blended with synthetics, we can put them into our garden compost piles or into large-scale municipal systems (if your compost provider accepts them). The composting process can be hastened if your materials are shredded or cut into small pieces. A German study that focused on making high-quality compost as a rural economic development strategy experimented with different ratios of wool and other feedstocks and determined that 25 percent wool, 50 percent green waste (yard clippings), and 25 percent manure made an excellent compost.

At the Fashion Institute of Technology, students took part in an innovative cotton muslin composting project. Muslin fabric is commonly used by students, designers, many brands, and the fashion industry as a whole to develop mock-up garments from their trial-run patterns. Much of the cotton used for these muslin pattern mock-ups is subsequently sent into the waste stream and eventually winds up in a landfill. The FIT students chose instead to compost their muslin with food waste to make soil for the school's rooftop dye garden. The project was a great success.

Composting not only keeps sewing scraps and clothing out of landfills and incinerators, preventing them from inundating the economies of foreign countries, but is also arguably the highest use for worn-out garments whose fibers are too broken down for repurposing or recycling. As a first step in this direction to generate the soil-to-soil textile economy, consider purchasing 100 percent natural fiber clothing, preferably undyed or naturally dyed so that you wear it safely. It has the highest chance of being recycled easily and eventually returned to the soil. If you already compost at home, try adding bits of your worn-out clothing to your pile. If you don't compost, we've seen success burying clothing items in flower or vegetable beds. During a garden-based experiment, we saw that within one growing season an entire generation of old organic cotton underwear was transformed by microbes into life-giving soil.

Laying your clothes out on hot, sunny days and rotating them to receive the sun's rays is a time-honored method for killing bacteria and cleaning fibers. If there is something on the surface of the garment that won't come off and necessitates washing, the delicate cycle in your washing machine or a good handwashing with a little soap and a warm-water rinse will do. Laying wool garments flat to dry helps retain their shape.

Thinking about how to keep our clothing's value by lengthening the amount of time we keep it in play, and then seeking out ways to recycle it

The compost system is one of the most recent sustainability initiatives that has been embraced by FIT. It was developed in part as a companion to the Rooftop Natural Dye Garden, where plants and flowers are grown, then harvested and dried to become a source of natural dyes. The compost project is helping to create nutrient-rich material so the dye garden can flourish. Here Amanda Farr (*left*) and Lydia Baird (*right*) dry compost on the FIT rooftop garden. *Photo by Smiljana Peros / FIT.*

into new yarns and textiles, is of course important. Yet we need more. We need to reimagine our whole approach to garment creation, from the farm—where clothing ought to be grown—to our skin and back to the soil again. Think about the history of our food system in this country. In the decades following World War Two, our food system became industrialized, favoring quantity over quality, heavy chemical use and synthetic agents over organic ones, environmental degradation over regenerative practices, genetically modified sources instead of natural ones, and short-term profit over long-term sustainability. In recent years, however, an alternative model of food production has arisen, based on the natural fertility of the soil and the time-tested agroecological practices of diverse cultures. The food revolution has swept the globe, transforming our attitudes toward this fundamental human need. Fiber is poised to go through a similar transformation. We do not need to accept the current model of industrial clothing production, with all its social, physical, and ecological costs. There is a better way.

A farm visit to Robin Lynde's Meridian Jacobs farm during lambing season documented the raw material source that provided the essential wool for the one-year wardrobe challenge. Kacy Dapp designed and knit the sweater-shirt, and the felted hat was designed and made by Katharine Jolda. Both artisans lived and worked in Oakland, a one-hour drive from Lynde's farm.

CHAPTER TWO

The Fibershed Movement

Similar to the organic and Slow Food movements, the Fibershed movement began small. In isolated pockets beginning in the United States, individuals and organizations began to envision an alternative to the inequitable business-as-usual model of garment manufacture and disposal. Some of the early leaders of this movement were motivated by the human and environmental toll being taken on their communities by the industrial clothing system; some were textile artisans who simply loved the feel and look of natural fibers and dyes; some were local economy organizers looking for ways to expand development opportunities for small businesses; some were environmental advocates concerned about energy use, carbon footprints, and climate change; and some were farmers and ranchers looking for venues to bring their products to the marketplace.

Over time these isolated pockets grew, as people began to beta-test their ideas and implement fiber and dye practices where they lived, often by linking with like-minded individuals, until a threshold was reached and a movement was born. There were many pathways to this threshold. In this chapter I will tell my personal story of how I came to the world of place-based fiber-and-dye systems, how my decision to do the wardrobe challenge year led to the founding of Fibershed, and how I subsequently worked to grow our work. I will offer lessons learned, gratitude expressed, and thoughts on where the movement is going from my experience building partnerships and advocating for fibershed systems. In subsequent chapters I will explore the new opportunities and challenges confronting the movement, as well as detail the nuts and bolts of making a fibershed function work.

Harvesting Color

My personal journey began during a summer teaching job in 1998 at the Arts and Crafts Center of the University of California–Davis. Inspired by an interest in textiles passed to me by a great-grandmother and mother who were talented seamstresses, I intended to hone my skills while also learning how hands-on education can be designed to empower the next generation. I had enrolled in a series of courses offered through a new interdisciplinary Nature and Culture program at the university, and that summer the center hired me to provide classroom education in textile arts for eight- and nine-year-old children, which included the dyeing of their garments. Prior to the start of class, I found myself planning my curriculum with the gloves, dust masks, and disposable aprons that I had been given by the center's staff for the textile-dyeing process. Each time the children and I worked on a textile project, we suited up in plastic before opening up our jars of colorful powder dyes and spooning heaping mounds of the chemicals into vats of water, being careful to not let the powders touch our skin or get into our eyes or lungs. After we finished dyeing the shirts, the students and I would pour the remaining dye down the drain.

As I cleaned the classroom after our group sessions, I began to wonder: Why were we doing this? Where had these powders come from? If we were prohibited from coming into contact with them during the dyeing process, why was it all right to pour the liquid residue down the drain and then wear the dye-soaked T-shirt next to our skin? I asked friends and fellow teachers what they knew about the chemical source of these dyes, receiving a great many *I don't know*s in response. Then the director of the Arts and Crafts Center shared a recently published master's thesis by Rachel Stone, a graduate of the university's design department. Titled *Opening Pandora's Box of Colors*, the thesis described the crude oil and coal tar refinement processes that form the molecular foundation for the synthetic colors I'd been using with the children.

After deepening my understanding of the implications of synthetic dye use, I decided to search for alternative substances that I could put to use in the classroom. A query on the internet (Ask Jeeves in those pre-Google days) provided a hopeful answer: plant-based dyes. Shortly I found myself at our local food co-op purchasing cabbage and beets, as well as collecting onion skins from the bottom of the onion bin. I rode my bike along the greenbelt to collect blackberries and dandelion leaves. At that time I had not read any natural dye books and had never been personally exposed in action or word to any natural dye processes. I decided

Different breeds of wool take up plant dye in subtle and different ways. Here handspun Corriedale cross and machine-spun merino yarns were dyed with black walnut.

I'd learn with the students; together we'd figure out how to cook our cabbage, beets, onion skins, and berries to make our own natural colors.

In class we dyed our clothing, ate the beets, and poured the residue water onto the lawn after it had cooled. The process was incredibly easy, and the curious students began to ask questions about every plant and tree around them. Soon we found ourselves walking across campus identifying species by name and gathering leaves and stems while conjuring up hypotheses on which plants we thought would yield what colors. Needless to say, the dust masks were put away, the plastic aprons were recycled, and we no longer had any use for the disposable gloves. Our textile arts class became the *ecology of color* course. A new curriculum was born that summer, and a new way of living with the natural world had been sparked for the children and myself.

I continued my natural dye experiments and began taking weaving courses at the university. A new question formed: How do I get deeply connected to color and form in a real and meaningful way? How do I get to the "backbone" of color? In the summer of 1999, a friend gave me a book by Canadian dyer Trudy Van Stralen that explained how to use plant and insect species from Europe, Central America, and South

America that have a long history as sources of natural dyes, including indigo, madder, and cochineal. Van Stralen also highlighted many local and easily accessible species such as marigolds and onion skins. Inspired by what I was reading, I began to replicate what I was learning and soon developed a hands-on approach to plant-based dye experimentation. During summer and spring school breaks, as well as after graduating from college, I saved money to support further study with ethnobotanists and artisans who had honed their craft in natural dyeing, including the teachers at Tierra Wools in Los Ojos, New Mexico; Michel Garcia in Provence; Carol Leigh in Missouri; Rose Dedman of the Navajo reservation; and Carol Lee in Wyoming.

In 2005 I traveled to Thailand, Vietnam, and Indonesia to learn from village-scale textile cooperatives that utilize a multitude of plant sources for dyeing, including morinda root, indigo, mango leaves, and plant species that have no English translation. While traveling through urban centers on my way into the surrounding hills and upland communities, I began to notice that many of the culverts carrying water and effluent were ripe with chemical contamination. I wasn't sure of the source until I walked across a bridge one day and saw a pipe with pale gray water pouring into a freshwater tributary from a manufacturing site where they were doing synthetic dye work.

While studying in villages near the Mekong River on the border between Thailand and Laos, I witnessed ceremonies held by community members who celebrated specific moments on the agricultural calendar by wearing symbolic garments that were in some cases the exact same ones worn two or three generations earlier honoring the same date. There had been no cultural push to acquire extraneous garments in the intervening years in these communities. The ceremony reflected the deep consideration that had gone into honing the right garment for the exact life experience a person was encountering. The creation and use of textiles that I experienced in these Thai villages was formative for me. It compelled me to reflect on how these values could be translated to my home region and involve a similar level of connection among the color, texture, form, use, and landscape itself.

> **The creation and use of textiles that I experienced in these Thai villages was formative for me.**

The cycle of learning and teaching is a continuous process in my life. Whatever I learn and synthesize, I quickly want to share widely with others (always with permission from those who taught me). So upon my return, I found myself developing a curriculum for a local museum in my longtime home, the Bay Area. I titled it Ecological Arts, and its purpose was to help children develop hands-on skills in the practice of growing and naturally dyeing their own clothes. Students analyzed every fiber and dye they used through a social and biological lens with the goal of answering key questions: Where does this plant grow? How is it harvested? Who processed it? For children, this understanding of textile culture was best translated through story and tactile experience; tales of the history of silk and tufts of wool were woven together, summer after summer.

As a young person growing into my own understanding of the world, teaching children such foundational human skills as spinning, felting, weaving, sewing, the respectful harvest of plants, and learning to care for their own natural dye gardens provided a window into our own anthropological evolution. I experienced the contrasts between the material culture of the past and the material culture of today. I'd show a child how to spin fiber from the native dogbane plant, a species that has been in use for millennia in our region and used for twine, string, and fish weirs. I would teach students to use a leg and hand to twist the fiber across their thigh, and I would watch as they had to adjust their slippery polyester pants to get the correct traction for the task. I taught them to hand-stitch a patch in order to mend a hole in their clothing, noting the irony that the child was dressed from head to toe in fossil-fuel-derived clothing produced eight thousand miles away by workers not much older than they were.

I watched my students grow into self-reliant, creative, and intellectually stimulated human beings. But a question also arose in my mind: Where in our culture could their new skills be practiced and advanced? Within their modern economy and culture, it was very likely that my students would not have the privilege of time to mend their own clothes, make their own handspun yarns, or work on a farm to further their understanding of the genetic diversity of our fiber system. Instead they would likely experience the pressure to keep up with the high-cost and high-volume expectations of modern living, becoming increasingly dependent upon the consumption of cheaply made things whose biological source (if there was one) would remain deliberately obscure. I wished my students would find a way to remain craft-focused, but the odds were stacked against them. I wondered how many young people would end up like me, searching for wild blackberries on a college campus to dye a T-shirt because they, and everyone around them, had forgotten they didn't actually need to use a costly and human-health-degrading carcinogen to color their clothes.

Rose Dedman of the Navajo Tribe cooks cliff rose, wild carrot, and ground lichen to make natural dyes for her wool yarns.

The Power of Place

From my hands-on learning journey that spanned from the Navajo reservation to the Thai village of Doi Tao, I learned about textile making
that was, for the most part, not reflected in today's industrial economy.
Instead I found geographic pockets of time-tested indigenous practices
and approaches toward textile making that existed symbiotically with
ecosystem health and function. In Northern California, for example,
there is a thirteen-thousand-year-old (some say longer) history of weaving with plant species such as willow, hazel, deer grass, sedge, and many
others. Baskets in this region are created for all types of function—water-
tight vessels for food, cradleboards to carry babies, and burden baskets
for harvest. Additionally many of these same plant species have also
been used in constructing garments.

With their intricately woven design and sturdy structure, the California baskets embody an ecological code of conduct. Harvesting for basketry,
when done well, honors the regenerating capacity of the plant species that
the weavers rely upon, thereby ensuring future generations of weavers an
opportunity to continue their textile traditions. I came to recognize the
indigenous understanding that plants are our relatives and deserve our
respect, whether it is through a weaver's singing and expressions of
thankfulness or through harvesting practices that enhance the regenerative growth cycle of the plant, including fire and *coppicing* (a process of
pruning the plant all the way to the base to allow for new growth). There
are many examples of ecologically driven, systems-thinking textile cultures all over the world, each honed by the geology, biology, and weather
patterns of their place. Through travel, workshops, and apprenticeships,
I have been able to gather knowledge and a set of practical skills that
were not readily available through my formal Western education.

I also learned I could harvest dye color from fewer plants if the soil
was healthy, with abundant microbes and fungi, proper levels of carbon
and other nutrients, a porous soil structure to hold water, and a diversity
of plant species on the surface. I learned that healthy soil requires a pro-
hibition on insecticides, herbicides, fungicides, and the application of
synthetic fertilizer, all of which damage or destroy the microbial life
essential to maintain plant vigor. In this sense growing plants for dyes
and textiles is exactly like growing plants for food—it is best to use an
organic, holistic system that restores and maintains healthy soils. The
co-benefits of this type of agriculture—whether for food, dyes, or fiber—
are substantial for human and animal health as well as climate change
mitigation and resilience. This includes the integration of livestock, such

as sheep and goats, into carbon farming systems. (I will cover this topic in detail in chapter 3.)

I came to appreciate and pay closer attention to the intricate ties between human culture—be it through language, medicine, art, religion, or family structures—and the influences of rainfall patterns, soil health, and biological diversity. From the complex double ikat weavings produced in the village of Tegenungan, Bali, to the intricate sedge and willow basketry of California's native tribes, communities that hone their relationship with their material culture within a strategic geography over hundreds of generations hold many powerful teachings for modern societies.

A picture began to emerge about how communities attempting to manage their natural resource base for long-term health had been repeatedly disrupted by external forces, often destructively. Foreign colonizers have reduced indigenous communities into economic units to be bought and sold, often achieving their goals through violence. It is an old story, as old as spices, silk, cochineal, indigo dyes, gold, cotton, and a strong human back to do the work, often in the form of slavery. Unfortunately the sanitized version of history that I was taught in the American school system was based upon a pro-industrial narrative marked by wars, treaties, traders, and the triumph of (Western) technology and progress over "barbarism" and ignorance—a narrative that normalized the act of conquest. Rarely did a history teacher drill down into the purely resource-driven motivations for such conquests.

But I re-educated myself. I read about the history of fiber-and-dye systems and began to understand how the lust for the deep-red dye provided by cochineal insects (among other treasures) drove Spaniards deep into Central America, and how the desire for cotton textiles drove the British to expand their empire into the Indian subcontinent and eventually into North America. These actions opened the door to a high level of cultural and physical violence that continues to ripple through our world. And today, after centuries of empire building that made these once rare and precious natural textiles and dyes accessible to mass markets, we are now inundated with synthetic colors and industrially grown materials and continue to fight wars and exploit communities in the pursuit of these resources, which we are overproducing on a mass scale.

The Wardrobe Challenge

While sitting in an airport in early 2009 on my way back from the Navajo reservation where I had been studying natural dye practice with several women and their family members, I watched the news and saw more American troops being deployed into Afghanistan. Feeling angry by the unending resource wars, I thought hard about what I could do as an educator and textile developer beyond signing petitions and speaking with my elected representatives. I looked down at the plastic chair I occupied and saw my gray stretchy corduroy pants, and I realized that both the chair and my clothing were made from the same raw material: oil. I looked at the carpet beneath my feet—it was also made of oil. The jet fuel that would power my ride home—oil. In fact, nearly everything surrounding me involved products that originated from fossil fuels extracted from war-torn lands. All this fossil carbon was then refined and burned to create these materials, in the process pumping significant concentrations of carbon dioxide into our atmosphere, creating a layer of heat-trapping gas and amplifying climate chaos. In 2009 any personal actions that generated alternatives to

> Nearly everything surrounding me involved products that originated from fossil fuels extracted from war-torn lands.

fossil carbon dependency felt like the most appropriate thing to do. To this day I remain confident in a "both-and" commitment to individual behavior change in tandem with support of large-scale intersectional and bipartisan climate justice movement-building efforts.

This commitment had been percolating for a few years, ever since I realized that while I was teaching people how to make textiles from natural fibers and dyes, I wasn't wearing what I was weaving and dyeing on my own body. I had a closet full of clothing largely sourced from fossil carbon fibers, and almost all were dyed with these same non-renewing materials, originating from deep within earth's ancient carbon pool. (I later came to understand that the manufacture of plastic-based fibers consumes more than three hundred million barrels of oil annually, while the textile industry's greenhouse gas emissions totals nearly one billion tons of carbon dioxide or its equivalent each year, more than all international flights and maritime shipping combined.) Sitting there in the airport that day, it became quite clear that I needed to do something different about the way I dressed myself. I needed to walk my own talk, interweaving my skills in processing natural fiber and natural dyes—and all the cultural and ecological integrity they represented—with my concerns about climate change, resource wars, and water pollution. By changing the way I clothed my body each day, I intended to unplug from my relationship with these unhealthy systems.

I made a decision: Upon returning home I would upend the way that I had been cultured to consume and wear textiles, and as a function of this change, I would allow myself and my body to become the center of an experiment. The airport experience was a lightning strike of sorts. I had a sudden image of a wardrobe that would be made from natural fibers and dyes grown within a strategic area centered on where I lived—a place-based geography. I later coined the term *fibershed* for this. I would focus my attention on the plants, animals, and people that lived in this fibershed with me and determine how, or if, I could create a prototype wardrobe and live in these homegrown clothing items for twelve months. It would be a test, focused upon how to create garments from the soils of the region where I lived.

I learned that in 1965, 95 percent of the clothes in the closet of a typical American were made within the political borders of our nation. Today that figure is only 2 percent. Unfortunately this offshoring of the textile industry was not prompted by a desire for higher standards of production, economic equity for workers, or tighter environmental regulation. Quite the opposite—it was done to circumvent the policies, unions, and environmental protection costs associated with doing business in America. During my teenage years and early twenties, I witnessed the lion's share of the death-stake thrust into the heart of American manufacturing. Policies such as NAFTA and CAFTA (the North American Free Trade Agreement and Central American Free Trade Agreement) became a perfect storm of federal decision making that enabled frictionless capital to quickly liquidate American assets while "reinvesting" in lower-wage nations. American manufacturing systems that processed non-perishable raw materials were much easier to move overseas—a fiber without an expiration date could travel anywhere to be turned into clothing items.

You might assume that as a consequence of this offshoring of manufacturing, the raw fiber material in my region would be in short supply. However, this wasn't the case. As I delved deeper into the resource base of my fibershed in an attempt to source materials for my locally grown wardrobe, I learned that there was an abundance of natural fiber and dye options available. Well known to many farmers and ranchers, though new to me, was the knowledge that there was a significant underutilization of available protein fibers (wool and alpaca mainly) in my region, largely because the fibers could not be easily processed locally. Meanwhile, many people in my community were jogging, climbing, and biking in wool base layers grown in New Zealand and manufactured in Asia. During this journey I discovered that California produces enough fine fiber every year to create several million garments and durable goods.

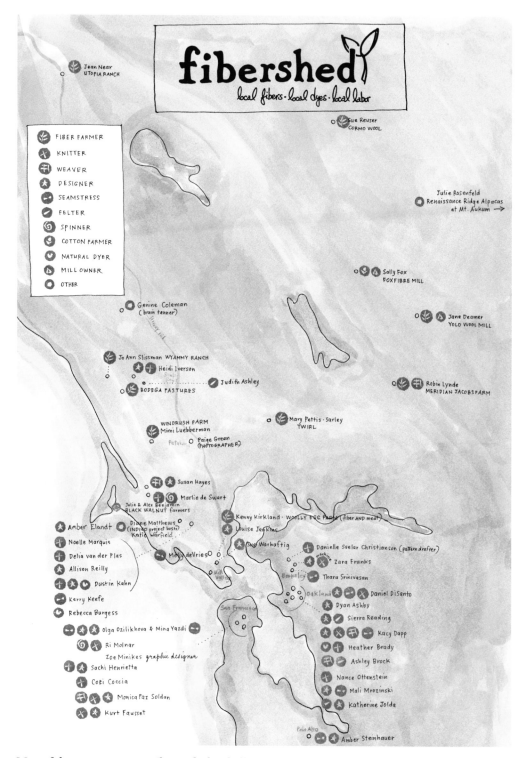

Map of the one-year, 150-mile wardrobe challenge. Each person who contributed is listed with an icon that shows their skill base. *Courtesy of Fibershed.*

With so much raw material readily available, it was both bizarre and frustrating to witness such high numbers of unemployed and underemployed people in my community who could be participating in meaningful work to add value to our raw materials.

Although the fibershed concept is based on a *watershed* or *foodshed*—a bounded geographic area where water or food is sourced—clothing is not as simple as food, which you can harvest raw, ferment, or cook over heat and put into your mouth and chew immediately. Creating a wardrobe involves materials that must be processed in multiple stages to become functional garments. I wondered: Did the equipment and skilled labor exist in my fibershed to make these steps possible? I knew there was a strong food community where I live that had developed over many years, but would it be possible to build a similar community for fiber?

I decided to commit to one year of sourcing all my clothing from my own fibershed, which I called the one-year wardrobe challenge. It began with a successful Kickstarter campaign to raise money to cover the costs of the raw materials and services of textile artisans, as well as to document the project's progress with longtime friend, colleague, and natural light photographer Paige Green. I researched, interviewed, and conducted the writing projects, which included articles about farmer-and-artisan collaborations. After the conclusion of the Kickstarter campaign, the next step quickly became apparent: It was time to set up meetings with neighboring farms, ranches, design students, and community members who were interested in collaborating on the development phase of garment creation.

Creating Community

One of the fibers grown within my fibershed is cotton, and I have a great appreciation for cotton fabric. I was keen to include this California-grown fiber within the wardrobe experiment. While I was well aware that cotton carried a set of complex ecological realities and impacts (discussed in chapter 5), I also knew that there is a wide spectrum of practices employed to produce the fiber. I quickly realized there is no one story you can tell about cotton unless you know the farm where it's being grown. I happened to be a two-hour drive from Sally Fox, a cotton breeder, textile designer, and farmer of organic, biodynamic, color-grown cotton, merino wool, wheat, and alfalfa. This close proximity provided me an opportunity to learn about the history of, as well as physically experience, the *Gossypium hirsutum* and *G. barbadense* species of cotton as a garment on my skin.

I also came to appreciate the fibers produced from the Camelidae animal family—alpacas, llamas, and guanacos—all of which existed in

Jean Near (*right*), the owner of Utopia Ranch, stands with California College of the Arts textile design student Sachi Henrietta (*left*) and Rebecca Burgess, admiring Sachi's knitwear skills.

my fibershed and provided delicately soft protein fibers. In the Sierra foothills, 147 miles from my front door, I had my first point of contact with Renaissance Ridge Farm, known for its beautifully colored and fine-fiber alpaca herd. Then while attending a county fair in Mendocino County, I met Jean Near, an inspiring female rancher over one hundred years of age, and came to know her flock of Merino sheep at Utopia Ranch in Redwood Valley.

Through my weekly trip to the local farmers market to purchase locally produced eggs, I discovered the owners of a wool-producing flock of sheep that was grazing in the midst of a suburb in Mill Valley, just across the Golden Gate Bridge from San Francisco. Kenny Kirkland's family homestead, founded in 1867, raises chickens, produces their own biodiesel, and maintains a small flock of sheep, all nestled on two acres among high-priced trophy houses that have since been built.

As I made fiber and yarn purchases from these farms and ranches, I simultaneously to set up meetings with friends and colleagues who were interested in making prototype garments from locally grown materials. Several local knitters kindly helped me make sweaters, hats, leg warmers, leggings, and socks, which would hopefully keep me warm in the most inclement weather. The approach of winter worried me because I wasn't sure I would have enough protection as the temperatures dropped and the rains arrived (even in California there is need for winter attire, although it would have been more intense in a colder climate). Wearing bulkier yarns meant layering several pieces onto my body to keep the winds from getting through the fabric structure. And while producing those layers took time, through the work of many individual knitters, just the right pieces came together to generate enough protection during the coldest months of the year. Eventually the prototype fibershed wardrobe grew to include shorts, wide-leg capris for biking, work pants, sleeveless tunics that double as dresses, separate sleeves, below-the-knee bloomers, socks, and undergarments.

As the project of growing, spinning, knitting, weaving, and sewing continued, it became clear that one of the greatest challenges in creating homegrown garments was the making of cloth. The mechanized machinery needed to manufacture fine-gauge yarns and finished textiles had not been available in my area for decades, and in some cases it had not existed for more than a century. As a result, we had to rely on human-powered equipment, including hand combs, spinning wheels, knitting needles, and floor looms, as well as the time-honored skills necessary for using the human-powered equipment. Fortunately we had the opportunity to work with the last remaining wool mill in our region, the Valley Oak Wool Mill. Its machinery had survived being melted down or sold to an offshore company thanks to the tenacity and heart of its original owner, Jane Deamer (the mill is operated today by long-time employee-turned-owner Marcail Williams). Jane had managed to remain in business with the support of local fiber producers, making yarns for the hand-knit and handwoven community. Her processing equipment provided the bulk of the yarns for my prototype wardrobe, and her machines continue to supply our region with homegrown yarns to this day.

One of the most important goals of the wardrobe year was to create community connectivity by matching a local artisan with a local farmer to build a garment. As the project grew over time, we identified our collective skill base and mapped the participating farms, ranches, spinners, weavers, sewers, natural dyers, and designers within the region. I was gratified to see small businesses begin to grow out of these experiments, as the people involved deepened their collaborations. Some of the fiber producers began to develop their own more complex blended farm yarns using not only material from their operations but also fiber purchased from their neighbors' and friends' farms. Many new yarn recipes emerged during this time. We were collectively learning and exploring the depth and boundary of the new arc of texture and color provided within our region.

In addition to the excitement of seeing a fiber, natural dye, and design community mapped geographically, I continued to experience the joy that came from observing fiber growers and designers discover one another for the first time. The project organically evolved into a social network of collaborators that spanned the soil-to-skin creation process. The quality

Marcail Williams, the owner and operator of the Valley Oak Wool Mill.

Less Is More

The Less Is More Wardrobe concept considers that less clutter is good for the mind and brings clarity to daily life; likewise, a simple, well-crafted wardrobe is about creating the conditions to support a greater sense of well-being. In our society many of us create a sense of self-identity and try to stimulate a sense of belonging through the clothes we wear, but all too often this means keeping our awareness of ourselves and others at the surface level instead of delving deeper into our truer human nature. If we construct and care for wardrobes that give us more connected, functional, and gratitude-filled experiences, then our garments are fulfilling their true mission. The idea of being warm when you need to be warm, cool when you need to be cool, protected from sun, and having clothing that moves well with your movements, clothing that honors a commemorative event—these are the fundamental reasons we have clothes.

A simple wardrobe allows you to invest in fewer, high-quality items that are well designed to meet your actual needs. When you slow down your rate of consumption, it is also possible to take the time to develop your own garments through sewing, knitting, visible mending, or felting. Here are four strategies to inspire and sustain a simple wardrobe:

1. Purchase items secondhand and adapt them through hemming lengths, overdyeing the clothing with kitchen-scrap dyes like avocado pits and onion skins, or sewing unique patches or trims that make the items stronger or more functional.
2. Purchase yarn (or cloth if it's available) direct from a farm or ranch (or knitting store that sells locally grown yarn and textiles), and knit or sew a garment that will function for multiple occasions.
3. Commission your own pieces with people in your community. There are farmers and designers who enjoy working with locally grown materials for custom knitting, sewing, and dye work. Building a small team to help you develop a meaningful garment and support right livelihoods is a great way to help build your fibershed economy and your wardrobe.
4. Consider purchasing from designers who produce timeless heirloom pieces that are able to biodegrade and eventually return to the soil. Seeking out supply chains that are as local as possible is of value to support meaningful livelihoods in our home communities.

of the craftsmanship that emerged from the collaborations of farmers and designers proved to be an unending wellspring of inspiration.

We continually surprised ourselves with our ability to devise local solutions when faced with challenges. In response to the need for specific dye colors, for example, we expanded an indigo- and coreopsis-growing project in the spring of 2010. These two species provided the design community with access to plant-based blue and orange. Other colors were sourced from naturalized and native plant species that we found throughout the gardens and ranches within the region. It was similar with fiber. According to convention, wool from meat sheep is not acceptable for clothing material, but we found this not to be the case in all circumstances; we made long underwear out of some of it.

It was profound to transform my anonymously made, industrially manufactured wardrobe to one grown, dyed, spun, knit, and sewn by people and animals completely familiar to me, head-to-toe. I learned quite a bit about hand-knit socks during this project, often having to darn a pair every three or four days. Corriedale wool from Mimi Luebbermann's Windrush Farm in the Chileno Valley proved by far the sturdiest wool for the job. Marlie de Swart, the owner of Black Mountain Weavers in Point Reyes Station, knit a mid-calf-height pair of socks that had been dyed in her black walnuts and my toyon (a native California chaparral species). To manage the rainy weather, it was clear that felt was going to offer the best option for my outerwear, and Mali Mrozinski and Katharine Jolda, both designers from Oakland, worked together on developing a felt coat from a black alpaca named Blackberry. Water sheeted off the coat in heavy rains; I could remain dry and not feel any form of chill from the wind.

The fibershed wardrobe was a vernacular that communicated the wild, naturalized, and domesticated diversity of my home region. The level of biological diversity that provided the color and form of my garments was all predicated upon the geology, rainfall patterns, slopes, and annual temperature fluxes that defined our bioregion. And just like in my fibershed, regional textile cultures around the world speak the visual language of the land and communicate the stories of its people. Altogether, these small but interconnected communities illuminate an opportunity and pathway to clothe ourselves in a manner that amplifies,

> The fibershed wardrobe was a vernacular that communicated the wild, naturalized, and domesticated diversity of my home region.

enhances, and ultimately harmonizes our personal needs with the greater ecology we depend on for self-renewal, balance, and resilience. A fibershed is not a new concept, but it's one we must return to and redefine.

The wardrobe challenge began as a rugged sensory experience. It didn't always feel flattering to my skin tones or body shape, and the effort broke down my previous sense of identity and my idea of what fashion is or how I wanted it to be. However, I soon realized that my clothing was telling a different kind of community narrative, one that gave everyone who wanted to contribute a means to do so. This wardrobe had little to

A visit to Renaissance Ridge Farm alpacas after the completion of the felted alpaca raincoat. Oakland-based designer Mali Mrozinski and felter Katharine Jolda collaborated on the piece and worked with fiber from Blackberry, the resident black alpaca whose fiber's color and strength provided an excellent source of material for this warm and water-repellent coat.

nothing to do with wealth concentration, fashion trends, or cultural dominance. The wardrobe was a living expression of where I live, what values I have, and what sort of community I belong to.

Rising from the Crash

In an interview with a Spanish journalist and filmmaking couple that took place during my wardrobe challenge year, I discussed our economy—which had taken a hard hit from the 2008 financial crash and was still reeling. This was an obvious example of how utterly disconnected our local economy was from any kind of regional material resource base and how easily a centralized economy could unilaterally destabilize so many communities. Yet despite the chaos of the crash, I told the interviewer that the sheep in my fibershed kept producing wool, the cotton plants kept blossoming, the natural dye plants kept growing, the knitting needles kept moving in people's hands, and the spinning wheels kept spinning with the weight of human feet. Nothing had functionally changed within the fibershed despite the domino effect of the mortgage-backed securities shenanigans.

Experiencing how everyday reality kept going on our farms, gardens, and looms, the daily radio and internet bombardment of financial chaos was put into a very different perspective. It showed me that the economy as popularly discussed in the media and experienced by most of us is a system that places value upon a discreet sliver of our actual shared experience. It also told me there is a great deal of potential to create a new and relevant currency or even a set of currencies that could underpin an economy that reflects real-time biological and human social factors. I experienced viscerally such a significant contrast between what I was hearing and what I was living in those years after the crash, and I remain convinced there are multiple economic models that could serve our lived experience and the ecosystems we rely upon with much greater levels of dignity.

As an example of disjointed relationships between supply and demand that have been set in place by the current design of our economy, I learned during the wardrobe challenge year that there was more raw fiber being produced regionally than our community could process into clothes. In the case of wool and other raw animal fibers, vast amounts of material sit in bags in barns while profit margins for farmers and ranchers remain razor-thin. Through a subsequent analysis, we discovered that California is processing less than 1 percent of the wool produced in the state. But we could easily ameliorate this imbalance between supply and demand through investments in regional infrastructure, including vertically integrated value chains. Why not grow, mill, sew, cook, eat, and wear as locally as possible? How is it that an item of clothing can travel fifteen thousand miles to a box store, dyed and stitched together with materials sourced from non-renewable fossil carbon resources, and still be cheaper to make than clothing from within one's own region? Why are most farmers and ranchers barely surviving economically and tossing their valuable wool into a ditch? Why is imperfect food thrown into a landfill? The level of poverty in our communities is a ripe arena for entrepreneurship. The fibershed wardrobe challenge brought these cultural and economic possibilities home to me in an unforgettable way.

The carbon emission savings and benefits of living this way would also be a boon for our regional economies. To further our understanding of the real value of regionally grown and processed garments, Thara Srinivasan, one of the fibershed designers, had a job conducting the scientific analysis for Peace Labs, a life cycle assessment group out of the East Bay. Thara discovered that the pair of pants she made for the one-year wardrobe challenge had *one-sixth* the carbon and water footprint of a conventionally produced equivalent. Of course, there are large and complex infrastructure hurdles to creating access to such garments for the general public, and we will need expert planning, business acumen, and engineering skills to overcome these challenges (this topic will be explored in greater depth in chapter 3).

The urban-rural connections made during t he wardrobe challenge year also changed the face of our fibershed's community. Four new formal businesses were started as a result of the project, as well as a suite of informal collaborations that continue to blossom today. But when the challenge year came to an end, it became crucial to continue the community we had built together and support its continued growth. We needed to figure out how to continue to support this community of designers and farmers to work together and further increase access to our local materials, thereby making this work more accessible to a larger swath of the population. This new level of awareness within my home fibershed was evidence that no biophysical limitation bars us from returning the natural fiber and color traditions to our closets. I realized that this re-engagement of an American textile tradition is waiting to happen, and it is up to our regional communities to define how it is accomplished.

The wardrobe challenge had no precise predetermined outcome when I started it, and in a sense it has never concluded; to this day I wear many of the handmade clothes created for the project. And because I wanted to continue providing educational opportunities to the public

The 2013 Fibershed Fashion Gala featured pieces made by designer collaborations. This jacket, made with Robin Lynde's handwoven Jacob wool fabric, was designed and made by Shari Schopp. The dress was designed and made by Meredith Buck and Marlie de Swart from Sally Fox's organic Foxfibre cotton. *Modeled by Lauren Ortega.*

about the abundant fiber resources in our home region, my collaborators and I decided a good next step was to develop an event that would put the wardrobe challenge clothing in front of an audience to help people see and experience what "wearing the landscape" looks like. To do this, we rented a feed barn near Point Reyes Station and used the straw bales to create a runway for models and seating for a general audience. We served wine that was grown and produced by local vintners, beer from our neighborhood brewery, cheese from our region's dairies, wild nettles harvested by friends, and a birria goat stew; my brother created a line of homemade kombucha to serve to our guests. The models wore pokeberry-derived, magenta-dyed wools, yarn dyed in madder root, and oak galls dangling from ears and around necklines. It was our intention to illuminate the intersection between our region's food and fiber systems, and we were delighted when a couple of hundred people joined us for the evening.

Thanks to a team effort, we were able to raise six thousand dollars that night, a sum that we intended to apply toward the costs of building a small cotton-processing mill for the region. However, the money raised represented only a tiny percent of the roughly five hundred thousand dollars needed to start a new mill, including the purchasing and importing of microspin machinery from India. As a result, our discussion shifted again to how to use our funds to leverage a wider and intentionally long-lasting impact in our area. The idea was floated to start a nonprofit organization to uphold a mission focused on local fiber and natural dye systems. We saw the positive role that local food systems organizers had in helping communities access locally grown, healthy foods, but there was nothing analogous in the fiber-and-dye system to support access to locally produced, clean, healthy textiles. We spoke with various experts, including pro bono lawyers, and soon the idea became reality. Fibershed (the nonprofit) was launched in spring 2012, with the educational mission to develop regional and climate benefiting fiber systems on behalf of independent working producers by expanding opportunities to implement carbon farming, forming catalytic foundations to rebuild regional manufacturing, and connecting end-users to farms and ranches through public education.

Wool and Fine Fiber Symposium

One of our first projects as an organization was the inauguration of a Wool and Fine Fiber Symposium we would continue to put on each fall. Its goal is to bring together people involved in producing "farm-fresh" clothing in the region, including agriculturalists, shearers, designers,

A free public wool-grading demonstration offered at Fibershed's annual Wool Symposium is led by Stephany Wilkes, sheep shearer, author, and president of the Northern California Fibershed Cooperative.

knitters, artisans, entrepreneurs, and the public. The event's goal is to facilitate community conversations around the soil-to-skin approach to garment production. The inaugural symposium quickly sold out, and the packed room was filled with inspiring conversation and energy. The nine fiber producers who made presentations that day covered a wide range of issues involved with sustainable agriculture, including holistic management of livestock, noxious weed control, herd health, water use, predator challenges, fiber quality, color, consumer demand, and milling shortages.

A direct outcome of the symposium has been the connections made between artisans and business owners who were in search of locally grown and dyed fibers and yarns and the farmers and ranchers who could supply them. One obstacle facing the design community continues to be a lack of transparency about the conditions under which fibers are produced. On the agriculture side an obstacle for many ranchers and farmers to selling locally was the lack of markets. At the first symposium a path through these obstacles began to emerge as relationships between producers and buyers were nurtured, demonstrating the critical role that face-to-face convening plays in broadening movements.

Another major endeavor for Fibershed at the time was the implementation of an online marketplace that offered customers everything

Twirl Yarns are created with natural fibers
dyed with both wild and domesticated plants
from Mary Pettis Sarley's Napa Valley ranch.

Agricultural Cooperatives

In 2018 the idea of an online marketplace was reborn as a service offered from the Northern California Fibershed Cooperative, an agricultural marketing co-op that promotes the use and production of regionally grown fiber. Through the co-op, farmers, ranchers, and other producers who create goods from locally grown materials can pool their resources, including money, brand appeal, and expertise and make their products available on the website to consumers. The goals of the member-owned and -operated cooperative include: supporting expansion into new markets for farmers and independent artisans; making it easier for people to find and buy regionally grown, domestically produced goods that benefit local land and economies; and providing training, workshops, agritourism projects, retreats, and other events that support livelihoods for fibershed producers.

Agricultural cooperatives have been around a long time; they have been a part of American agriculture for more than one hundred years. According to the USDA, there are three thousand agricultural cooperatives in the nation with nearly three million members doing ninety-six billion dollars of business annually. Cooperatives are an efficient way for individuals to work together collaboratively, and they represent every aspect of agriculture, including growing, supplying, processing, packing, and marketing. They are structured organizations with bylaws, democratically controlled by their memberships, and recognized by federal and state governments as "useful to society"—which means they are exempt from anti-trust laws. They have numerous tax benefits and exist to serve their members' needs, principally by pooling resources and working together toward defined goals. Profits are distributed evenly to every member. The inclusive "one member / one vote" nature of cooperatives means they cannot be controlled by a minority of people and can resist the undue influence of wealthy investors.

from raw fibers to finished goods. The goal was to host an internet-based venue for individual farmers and ranchers to offer their fiber, yarns, and batting directly to the community, giving artisans, knitters, store owners, and others the ability to source raw materials with ease. If someone living in San Francisco didn't want to drive to Napa Valley for yarn, the farmer or rancher could send them the product. It also created an opportunity for online customers to buy finished garments, obtain seeds for dye gardens, or sign up for workshops and other educational activities. For us at Fibershed, the products sold through the marketplace helped illuminate the collaborations taking place within the local community.

The project was an initial success, signaling the need for an online market, but our lawyers said it looked too much like a business and thus exceeded the scope of Fibershed's status as a nonprofit organization, so we set the project aside to focus on our educational and land-based activities. It took us four years to find a pathway to bring the marketplace to fruition once again under a different legal architecture. It now exists under the umbrella of the Northern California Fibershed Cooperative, a producer-led agricultural cooperative.

Producer Program

Another project coming out of my wardrobe challenge year was the implementation of a formal artisan and producer sign-up program that has today over 170 members in our fibershed. The goal of this program was to let the community know they can buy directly from independent businesses that support regional agriculture, a fair economy, and ecological balance. This producer program, which has a strong online presence on the Fibershed website, has catalyzed many new and direct connections between urban and rural community residents and particularly among designers, students, farmers, and ranchers. All members are required to produce their fiber and natural dye within fifty-one counties in Northern California, with exceptions made for cotton and fine-gauge wool milling, which continues to go out of state due to a lack of milling capacity, and sewing thread, which is not yet manufactured in our region. In exchange, members are included in Fibershed's online directory; have the option to receive soil carbon baseline testing for their farms; are given certification tags, which they can place onto their products for sale; and have access to a free photoshoot and story designed to feature their life and work.

Fibershed producer meet-up held at Carleen Weirauch's sheep farm and creamery in Sonoma County. The producers meet twice a year to discuss their projects, interests, and collaborations.

Recently the boundaries between the producer and end-user categories have become blurred, as more designers are spending time on the participating farms—picking cotton, shearing sheep, and harvesting indigo—while many farmers are testing the waters of modern garment design themselves. As it evolves, the producer program continues to function as a foundational service to grow new small business ventures in the area. The ripple effect has been tremendous. Many interconnected pathways have emerged as the result of the commitment to wear locally grown, dyed, and made garments. The potentials for more continue to blossom each year, grounded in knowledge about how rich and life affirming our work can be. It also is a testament to the power of community and what can be accomplished when people work together to create an equitable, healthy, and beautiful fiber and food system.

As a result of these Fibershed projects, it became evident how essential our public education efforts were to the burgeoning fiber movement. The general populace had largely forgotten that clothing was an agricultural product, and many people were completely unaware of the high levels of toxic chemicals or the social and environmental costs involved in the creation of their clothing. Many potential customers were also unaware that they had an alternative in natural dyes and fiber sourced locally. For farmers and ranchers, there was an educational curve, too. In addition to learning about and implementing new climate benefitting practices, many producers who sold directly to the public were using synthetic dyes to achieve the bright colors they thought customers wanted. We encouraged producers to take a risk and invest in the arc of texture and color of our region with more natural dyes—and customers magically appeared.

> It became evident how essential our public education efforts were to the burgeoning fiber movement.

Early Fibershed Pioneers

I was blessed to meet and work with a group of inspiring renaissance women in these crucial early years, who all had distinct specialties in the fiber community. These relationships were formed through our collective interests and mutually supportive skills and services, and as the wardrobe challenge merged into Fibershed the nonprofit, they became incredibly important to our work, especially as we began to tackle the links between local-scale fiber enterprises and global issues, such as climate change. I'd like to share a sampling of the many stories from a small subset of the larger group of women who contributed to the wardrobe—women who were, and continue to be, an inspiration, representing a cross section of the fibershed model.

Monica Paz Soldan was one of Fibershed's original artisan members. A lifelong knitter and weaver, she has extensive fiber farming experience, having spent her formative years on her Danish grandfather's cotton and dairy farm in Arizona, where they grew a variety of cotton called Pima. Ironically, this variety originates in Peru—the home of another of Monica's grandparents. As a young woman, she felt the tug of fiber in her life but also the conflicts associated with its industrial production. Monica's Danish grandfather farmed cotton as a monoculture crop, embracing the "chemicalization" of the industry, as she put it. A common sight in her childhood was water in an irrigation ditch near the farm, flowing with a strange fluorescent green color. She lost an uncle to cancer, likely as a result of pesticide poisoning. On her Peruvian side, however, was a five-thousand-year tradition of sustainable cotton cultivation, possibly the longest on the planet. Cotton in this region came in a range of colors naturally, including red, green, and brown. Most of the fiber is still processed by hand and has not been treated as a commodity.

Monica Paz Soldan, 150-mile wardrobe challenge textile designer and knitter, works with color-grown cotton from Sally Fox's farm.

After a move to San Francisco, Monica began a home business called Tiny Textiles. Employing sewing and cutting techniques she learned from her Welsh grandmother, she sold knitted textiles first to friends before expanding her reach by word of mouth. I met Monica at one of the early presentations I did in preparation for the wardrobe challenge year, and she enthusiastically contributed her immense skills to the project. She is well known for her intricate designs and stitchwork, creating patterns that "make the fit joyful" in her words. Her layered designs for the garments she created allowed them to stretch comfortably, and the dyes

Hazel Flett of Bodega Pastures in Sonoma County stands alongside raw and freshly shorn wool that is stored in the barn she and others use to teach local schoolchildren about local fiber.

came from her backyard: fermented and fresh indigo, coreopsis, pokeberry, madder, and weld. She likened her dyeing process to abstract painting, a kind of purposeful imperfection that included blended and marbled colors in the yarns. For Monica, there is great joy in handmade goods. In that sense each Tiny Textiles piece embodied the fibershed model, from the soil of the pastured wool to the seeds of an urban dye garden and the spirit of the artist herself. Although Monica has since moved out of our Fibershed to start a permaculture-designed natural dye and food farm in eastern Washington, I will forever be thankful for the kindness, generosity, and incredible skill that infused her work and attitude.

Hazel Flett, working within a community of thoughtful land stewards, has lived and worked at Bodega Pastures, a beautiful eleven-hundred-acre sheep ranch located near the coast in western Sonoma County, for over thirty years. She knows every animal on the farm personally and raises a flock of mixed-breed sheep, including Suffolk, Wensleydale, Romney, Corriedale, and Navajo Churro. On her daily walks Hazel keeps a record of each sheep, working hard to memorize not only their outward appearance but also their behavior, mothering ability, fleece quality, and reproductive capacity. A native of England, she took to knitting as a child and has been involved in fiber arts ever since. It was a natural leap for her to become a wool producer when she joined Bodega Pastures. Together with her community members, the ranch produces nearly one thousand pounds of woolen fiber from the flock annually, which they sell as yarn, batts, fleece, and other products. Bodega manages their sheep herd in alignment with nature's model of herbivory. The flock is moved frequently through the pastures to avoid overgrazing and to maintain healthy forage, no synthetic chemicals are used, and the animals are treated humanely. Bodega wool is beloved among the fibershed community for its color and texture.

Katharine Jolda is one of Bodega Pastures' artisan supporters, a self-trained felter who grew up in Oakland and spent six years working on the Navajo reservation in Arizona, primarily with sheepherding and farming families. She was an early and enthusiastic participant in the wardrobe challenge year, creating incredibly durable felt garments for me, including a beautiful fedora and handbag made from Bodega Pastures wool. Her work is known as much for its innovative design as its utility and durability. Her incredibly warm and comfortable "147-mile" winter coat for the wardrobe was her first attempt at felting alpaca. Katherine is also committed to the reciprocal relationships at the heart of the fibershed model. She is a founder of the Yerba Buena Institute, which teaches craft-based ecological knowledge of the Bay Area, and a member of Black Mesa Weavers for Life and Land, which works to bring together cosmopolitan and indigenous fiber artisans.

Katharine's felting process is representative of her years of participation with the Navajo culture, which has lived within its ecological means for centuries. Born of water, wool, soap, and human hands, felt is the most immediate textile that can be created from the back of a sheep. Given the directness of the process and the practicality of the finished product, felting has found its way deep into her heart and soul and inspired many who have come into contact with Katharine. She's also a natural inventor; to process the fiber once it is shorn from the sheep and after it is washed on the reservation, she devised a bicycle-powered carder to comb the wool, which she dubbed the Cyclocarder. This human-powered drum carder creates on average six ounces of combed wool in only four to five minutes. She brought the Cyclocarder back home with her to Oakland, where it has made quite an impact, especially since it fit so well with the area's bicycle culture. Katherine's inventiveness was born of necessity and is now being incorporated into a community that for all its financial wealth has little access to equipment and manufacturing. The drum carder is as needed on our farms and studios as it was on the reservation.

Katharine Jolda's human-powered
bicycle drum carder at work—
designed by Katharine while she was
providing health education on the
Navajo Nation and helping with
shepherding and farming.

Robin Lynde raises an heirloom breed of Jacob sheep on her Solano County farm—Meridian Jacobs. Lynde has a farm store, offers classes in a range of fiber skills, and makes an impressive farm yarn that was used by Kacy Dapp, a talented Oakland knitter, to make the "Sweater Shirt." This vital item of clothing was worn by Rebecca Burgess as a next-to-skin garment throughout the winter and spring of the 150-mile wardrobe challenge.

Robin Lynde is a sheep farmer and weaver who lives on a ten-acre farm called Meridian Jacobs in Solano County, on the western edge of the Sacramento Valley. Her flock size is sixty-five head and primarily consists of Jacob sheep, a heritage breed known for its black-and-white fleece, multiple horns, and high-quality fiber. She also owns a few Blue-faced Leicester sheep, which when crossbred with the Jacob can add luster to the soft woolen fiber. Robin does it all. She selects and sorts the wool for spinning, creating unique yarns from each year's wool crop; she weaves blankets on a dobby loom; she stocks spinning and weaving equipment; she creates on-farm commercial yarns, handwoven pieces, and gift items; she teaches classes in weaving, spinning, and dyeing; and she hosts open house events on her farm three times per year. Recently she created the popular Farm Club for people who want to know more about what it takes to raise sheep and produce fiber. To keep the farm financially sustainable, Lynde sells Jacob lambs, known for their mild-flavored meat, direct to Farm Club members. It all adds up to an intensely personal enterprise that she eagerly shares with the wider world, creating strong, mutual relationships that have contributed significantly to our fibershed.

At the start of my wardrobe challenge year, I had met Sally Fox, a color cotton breeder as well as a sheep and grain farmer. Her farm is certified organic and biodynamic. Thirty years of plant breeding by hand has enabled Sally to cultivate varietals of color-grown cotton, including Buffalo Brown, Coyote Brown, and Palo Verde. Her plant breeding is focused on classic methods, requiring meticulous skill, patience, and fieldwork. As a result, she has developed seeds that produce colorful fiber that has enough length to be manufactured on mechanical cotton-spinning equipment. This revolutionary advancement brought opportunities in the early 1990s for brands like Levi's, L.L. Bean, and others to purchase Sally's cotton to include in the production of their organic denim and shirting material, which requires 70 percent less dye than garments made from conventional white cotton. For a while jeans were being made and grown in America originating from organically farmed soils. However, when the North American Free Trade Agreement (NAFTA) came into being, the first textile mills across the nation to shut down were the dye houses, which of course are an integral aspect of the textile production system. As a consequence, partner mills that carded, spun, knit, and wove textiles shut down in rapid sequence, resulting in much less demand for Sally's cotton despite its utility. Fortunately, she was able to tap into the burgeoning fiber movement, selling to smaller brands, handweavers, and spinners. Her knowledge covers every facet of color cotton production, from its science to its cultural history.

> When the North American Free Trade Agreement (NAFTA) came into being, the first textile mills across the nation to shut down were the dye houses, which of course are an integral aspect of the textile production system.

Sally purchased a small flock of colored Merino sheep to help control a noxious weed infestation and create a commercially viable blend of cotton and wool to sell. Today she also manages the 140-head flock for carbon farming purposes with the goal of increasing the organic carbon in her soils, which she has doubled in a decade. She embraced Sonora wheat for the same reason. She also loved its taste. This heirloom species, one of the oldest varieties in North America and popular as a traditional source of flour for tortillas, has exceptionally long roots, which means it can send carbon deep into the

The Backyard Hoodie

In 2014 cotton growers from our fibershed community were highlighted and the fibers from local farms utilized by The North Face in its Backyard Hoodie project. The company was headquartered in Alameda, California, at the time, which meant its location was a great foundation to spearhead its own 150-mile garment challenge. Its effort to bring a bioregional garment to the marketplace was the first attempt by a large-scale brand since NAFTA. The North Face is owned by VF Corporation, the largest textile conglomerate on Earth; it was a refreshing surprise to work at such an innovative level with such a significantly sized entity. This hoodie highlighted a culturally and

The North Face Backyard Hoodie—a 150-mile challenge to make a garment from within the company's fibershed—is made of cotton from Sally Fox's Viriditas Farm and the Sustainable Cotton Project.

economically transformative new ethos that not only positively impacted the farmers (who received new markets for their cotton) and wearers (who received an opportunity to invest in their local and conscious farming community) but also had a transformative impact upon the people within the company.

We watched a creative vision blossom from the designers to the sustainability and public relations teams. The commitment to work with materials that were grown within the region and not use synthetic dyes was a healthy challenge that stretched what people at The North Face knew to be possible. The company sourced its raw color-grown cotton from Sally Fox and a white Pima from the Martin family, who are members of the Sustainable Cotton Project, a group of family farmers in California's Central Valley growing cotton using biologically based pest management practices. The North Face also made every effort to bring manufacturing home from overseas; the growing and sewing took place within fifty miles of the company's headquarters, and the pattern was based on a very-close-to-zero-fallout design by technical designer Daniel DiSanto.

However, because there were no mills within our region that could do this work, the carding, spinning, and knitting took place in North and South Carolina at Hill Spinning and Clover Knits. Nevertheless, everyone was pleased to see that the entire garment was manufactured within the United States and that the company committed to doing everything physically possible within a half-day drive of their headquarters. The hoodies proved to be popular in the marketplace, too. The limited edition run sold out quickly even at a retail price of $125, double what conventionally produced hoodies cost. This project has led to continued positive moves by the company to source regionally grown natural fibers and to consider tightening supply chains to sustain livelihoods both regionally and domestically.

Thara Srinivasan, PhD, is a self-taught sewing-pattern developer who designed and constructed a pair of pants from Sally Fox's buffalo brown cotton flannel for the one-year wardrobe challenge. Srinivasan worked for the Pacific Ecoinformatics and Computational Ecology Lab in Berkeley, California. She conducted the first life cycle assessment for the one-year wardrobe challenge. Her comprehensive research showed that the pants she constructed had one-sixth of the carbon footprint of conventional alternatives.

soil in exchange for nutrients. She began with a few acres of Sonora wheat and today, that has expanded to almost thirty acres. After nearly disappearing, the species was reintroduced in California by Monica Spiller, a seed saver and founder of the Whole Grain Connection, which is dedicated to reviving stone-ground milling in the state.

In the span of a few short years, what had started as a personal experiment to see if I could thrive in a locally sourced wardrobe of natural fibers and dyes has blossomed into a hopeful, community-wide endeavor to revive and expand natural fiber and dye textile production in our fibershed. We have since become aware of similar work in other regions, and a great deal of cross-pollination has begun to take place among numerous farmers, ranchers, textile designers, artisans, clothing manufacturers, and others around the possibilities and opportunities embedded in the rapidly growing fibershed model. At the same time we are more and more aware of the links between natural fiber production and global threats, including climate change, and the ways that fibersheds throughout the world can be a major part of the solution.

Marie Hoff leads Ouessant sheep to a new pasture, the movement of the flock enhancing their service as strategic grazers that are cycling nutrients and reducing the fuel load to lower the risk of fire. *Photo by Alycia Lang.*

CHAPTER THREE

Soil-to-Soil Clothing and the Carbon Cycle

ature is full of cycles. Plants and animals are born, grow, reproduce, and die, returning to the earth from which all life springs. Water travels in cycles from sea to clouds, then to land, lakes, creeks, and eventually back to the sea again. The carbon cycle moves molecules through and between ancient carbon pools on the planet, including oceans, plants, soils, and the atmosphere. One vital carbon flow in particular is the movement of carbon dioxide from the air into a plant leaf, then to roots and soil, before a portion returns to the sky. Nature cycles and recycles, returning organic material to its source—the soil—to start again. Decomposition is nature's way of turning expired life into new life; nothing is lost, nothing pollutes, nothing is thrown away. Every part of nature has a "full accounting system" that balances income and expenditures over time even if there is an unexpected disruption, such as a forest fire or a hurricane. That's the beauty of self-renewing cycles—they replenish nature's checking and savings accounts.

Imagine for a moment a free-ranging airborne carbon dioxide (CO_2) molecule being absorbed by the green leaf of a plant. Once there, powered by the energy of the sun, the process of photosynthesis separates the carbon molecule from the two oxygen molecules, which are subsequently released back into the air for us to breathe. The carbon is transformed into simple carbohydrates (carbon + water) including two sugars (sucrose and glucose). The plant uses carbohydrates to feed complex soil ecosystems through its roots as well as to build its very structure. Imagine the plant as one of many in a grassy meadow, which is 40 to 60 percent carbon by weight—all of it captured from the atmosphere. Now picture a hungry herbivore consuming the plants in the meadow, turning the carbon into various types of protein, including the animal's coat—wool, for example. The wool is then shorn, processed, and dyed with natural colors, thus becoming yarn. Soon it is made into a sweater. After many years of repeated use, the sweater containing that original footloose carbon molecule ends up in a compost pile along with many other types of organic matter, gently decomposing. Eventually the finished compost finds its way to a garden, farm, or rangeland where it feeds the soil microbes that help to maintain a healthy plant community, and so begins the cycle again.

Now consider a second free-ranging airborne carbon dioxide molecule floating in the air. It is pulled into a green leaf as well, and the carbon is incorporated into sucrose, but this time the sugar molecule makes its way to the roots, where it is pushed into the soil by the plant as a part of its symbiotic relationship with mycorrhizal fungi and other soil microbes that help to produce the substance called *humus*. If you're a gardener, you know all about humus: It is the dark, rich, life-promoting soil essential to growing healthy plants. If left undisturbed—not tilled or consumed

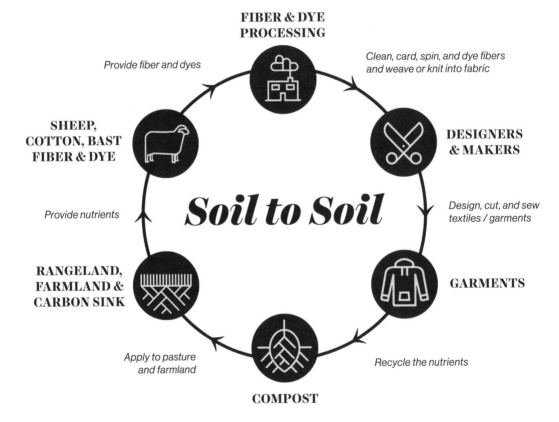

The Soil-to-Soil concept at the heart of the fibershed model. *Illustration by Andrew Plotsky, courtesy of Fibershed.*

by hungry microbes—the carbon stored in humus can stay safely sequestered in the soil for long periods of time. Meanwhile a vigorous "barter
economy" is taking place underground between plant roots and soil
microbes, exchanging carbon for essential nutrients that green plants
need for thriving, which makes the plant attractive to a hungry herbivore, which consumes the leaves containing other carbon molecules,
sending everything round and round once more.

The soil is one of the five great natural carbon pools, or sinks, on the
planet. The other four are the oceans, the atmosphere, the lithosphere
(the earth's rocky crust), and the biosphere (where humans live). The
planetary ebb and flow among these five carbon pools over short and
long periods of time forms the essence of the natural carbon cycle. In
modern times humans have become heavily dependent on three-hun-
dred-million-year-old carbon trapped in the lithosphere in the form of
coal, oil, and natural gas deposits to run our industry and transportation
sectors. However, continuous burning of these energy sources for more
than a century through engines, factories, and power plants has produced so much additional carbon dioxide that it has overwhelmed the
atmospheric carbon pool, causing climate chaos. Green plants do absorb
carbon being pumped into the sky, but they alone cannot balance the
carbon account. Soil needs to make up the difference. Organic and eco-
agricultural models of food production and land management can help
achieve this goal.

Considering Clothing from Soil to Soil

When we consider our clothing choices, we often make decisions based
on color, size, and style, not on the carbon pool from which a piece of
clothing originates. But the ultimate source of your clothes makes a huge
difference. We should be considering clothing in the context of the following questions: Are we wearing clothing from the fossil carbon pool or
are we wearing clothing grown in the soil? Are our clothes designed to
return to the carbon pool from which they came? Can we compost our
clothes and return them to the soil? Does our clothing degrade into
smaller and smaller synthetic fossil carbon fibers (plastic) and create
microfiber pollution in our oceans, soils, and landfills? How can our
clothing choices help us move atmospheric carbon into the soil? Then we
must discontinue our part in the production of carbon dioxide emissions
by divesting heavily from fossil carbon sources of clothing: virgin acrylic,
nylon, and polyester, as well as synthetic dyes and synthetic finishing
agents. Once we regain our focus on natural fiber systems—materials

that are farmed, ranched, and in some rare cases wild-harvested—the opportunity to restore carbon to our soils becomes a reality.

The next step is to restore and maintain healthy soils on our farms and ranches. Again, how we grow organic food is a good comparison. The same factors that make the soil productive for organic food—abundant microbial life, proper nutrient cycling, vigorous plants, holistically managed livestock, and resilient watersheds—also produce durable and self-renewing fiber for our clothes. Plant and animal fibers have historically been grown without pesticides, insecticides, or fossil fuels, and instead *with* natural processes that mimic natural ecology. The Soil-to-Soil system is the same whether a farmer is producing organic lettuce or organic cotton, and consumers should be just as demanding when considering the source of their clothing as they are about the source of their salads. This natural affinity between food and clothing, as well as their corresponding markets and social movements, is a critical element in the fibershed model.

> **Consumers should be just as demanding when considering the source of their clothing as they are about the source of their salads.**

Since the start of the Industrial Revolution, scientists estimate that we've lost 136 gigatons (136 billion metric tons) of carbon from our soils globally, and the excess carbon dioxide in our atmosphere from soil degradation and fossil carbon burning is trapping the sun's long-wave radiation and heating our planet at a pace unprecedented in geologic history. Making matters worse, our industrial fiber and food systems are net emitters of greenhouse gases, contributing approximately 50 percent of annual global emissions.[1] Nearly every aspect of the production and distribution of food and fiber in industrialized nations, not to mention the waste they generate, carries a large carbon footprint, including the fossil fuels required to power machinery and transportation, as well as the synthetic chemicals used as fertilizer and herbicides to keep yields up.

However, as dire as it sounds, this situation can be turned around. Soil-to-Soil systems are designed so that agriculture and its supply chains can move from being net emitters of greenhouse gases to net *reducers*. In other words, by employing region-appropriate carbon farming practices, food and fiber systems can help build up the soil carbon pool over time and thus become part of the solution to climate change.[2] This turnaround in food and fiber systems is good for farmers and ranchers as well, because global warming is creating increasingly frequent and unpredictable weather extremes, making it more difficult to farm and ranch across a wide variety of landscapes. We are seeing an increase in soil moisture loss due to evaporation and an increasing rate of soil erosion due to severe weather events, from floods to droughts. The overall heating of our planet is also contributing to carbon losses in our soils.[3] Altogether, these outcomes of climate change are leading to wide-spread soil degradation. There is a great need to create resiliency strategies that can support farmers and ranchers in growing and producing our fiber and food regeneratively.

Building soil organic matter has an important additional benefit: It increases the soil's water-holding capacity, thus reducing agriculture's need to draw from aquifers and surface water sources. This means water can be conserved for other uses, including the survival of imperiled species, lessening the tension between biological diversity and human-managed systems. From an economic perspective, improving the soil's ability to capture and retain water can be done naturally and relatively inexpensively while increasing yields and net primary productivity on working lands. This equates to increased fiber and food security and less land required to meet our essential needs. Focusing on the ground beneath our feet is a place-based strategy that every community can engage in without the need for complex technologies that come with hidden costs.

An illustration of the many benefits that accrue when soil organic matter is increased comes from research conducted by the Silver Lab at the University of California–Berkeley. In 2008 a single half-inch layer of compost produced from green waste and animal manure was applied to two different types of grazed rangelands in Northern California to see whether there would be any increases in forage production, soil carbon, and soil water-holding capacity and whether these increases could be sustained over a period of time. Initial results were very promising. After three years data showed that forage production increased by 40 and 70 percent, respectively, on the two sites; water-holding capacity of the soils increased by nearly 25 percent, while soil carbon increased by about 0.4 metric ton (the equivalent of 1.49 metric tons of CO_2) per acre per year. Significantly, these increases have persisted across ten years of data collection, and ecosystem modeling by the scientists at the Silver Lab suggests that these improvements will continue well into the future—all in response to a single application of compost.[4] The research provided an important starting point for proving, through peer-reviewed science, what the ecosystems response to adding energy (carbon) into the system could do to regenerate that system while having a positive climate impact. This body of research enabled changes in policy at the state level.

Water and California's Alexander Valley

Fresh water is a scarce resource, making up less than 3 percent of all water on the planet. Under climate change, droughts are predicted to become more severe, which means managing our freshwater resources is an increasingly vital task. We need every drop we can find. One new approach being taken by state and federal agencies is to study the volume of water stored in a body of *soil* (as opposed to a lake) and determine what practices need to be implemented in order to increase the amount of water captured there.

For example, a recent United States Geological Survey (USGS) assessment using a water balance model, which analyzes the correlation between the quantity of organic matter in a body of soil and its potential capacity to retain water, focused on the Alexander Valley, a part of the Russian River watershed in Central California. The study looked at how this correlation translated to the region's ability to offset a significant portion of its Climatic Water Deficit (CWD)—an important but technical term that describes a projected deficit of water availability as a result of increased heat due to climate change along with increased demand from agricultural and municipal water users. As temperatures warm, evaporation rates are expected to rise, which will increase the CWD.

Within the Alexander Valley, the estimated demand for irrigation water is projected to increase by two thousand acre-feet (650 million gallons or nearly eighty-seven million cubic feet) by the end of the century under a business-as-usual scenario. If the agricultural lands in this region were to increase their soil organic matter levels by 1 to 2 percent, however, the water-holding capacity of that soil could increase by as much as 25 percent. This has the potential to reduce the CWD of agricultural systems in the valley by up to 776 acre-feet (thirty-four million cubic feet), or 30 percent. "Water that stays in the watershed will also serve to preserve baseflows and riparian systems during low-flow periods and can potentially serve to sustain infiltration to the groundwater system," said Dr. Lorraine Flint, who with Dr. Alan Flint is a hydrologist with the USGS.[5] The two recently developed a comprehensive analysis that mapped and measured the soil water-holding capacity impacts related to soil organic matter increases in all of California's soils—they noted that increases in total soil organic matter of 3 percent in California's working lands would increase water-holding capacity by 4.7 million acre-feet. (Every year, the soils could capture this much more additional rainfall in the wet season and hold it to recharge aquifers, increase surface water flows, and sustain life. This is the amount of water used by over 12.5 million California households per year.)[6]

The more we learn about the positive impacts of increasing organic carbon levels in our soils, the more promising the opportunities become when we consider the potentials of scaling the work across broad landscapes. This analysis also puts into question and allows us to develop a more nuanced approach to traditional agricultural water budgets for crops and rangeland grazing. If cropland is managed to become a water sponge that recharges aquifers at rates higher than what is consumed to grow that crop, there is a real possibility for growing fiber, food, and dye in a manner that regenerates hydrologic function in our farm and ranchland.

Wild Garden Farm in California's Chileno Valley hosts a Fibershed carbon farm cohort meeting with neighboring fiber and dye farmers to assess the landscape for a hedgerow planting. The farm is working on building above- and belowground carbon to enhance productivity and water infiltration.

There are many methods for increasing soil organic matter (which is roughly 58 percent carbon), of which compost is just one. Integrated crop-and-livestock systems, windbreaks, hedgerows, silvopastures, and riparian corridor restoration are a few noteworthy practices that are commonly implemented in our region. These methods are most effective when stacked and combined into a farm or ranch setting. Regardless of the methods a land manager chooses for their particular place, the Soil-to-Soil system provides organizing principles that are key to regenerating soil, food, and fiber.

Restoring Our Rangelands

Rangelands have considerable impact on the movement and flow of carbon on our planet. Forty percent of the earth's land cover is in semi-arid and semi-humid zones that co-evolved with grazing animals beginning forty million years ago.[7] Because of their vast geographic extent, properly grazed rangelands have great potential to function as a significant sink for atmospheric carbon dioxide. By some estimates these rangeland systems could hold as much as 25 percent of the potential carbon sequestration of all global soils.[8]

Properly scaled in space and time, herbivore grazing stimulates plant growth through a variety of mechanisms resulting in increased root activity, which in turn can increase carbon capture. The earth's large herds move across the land on migratory schedules and are socially organized to remain together. Constant movement also means herds can avoid grass that has been temporarily soiled by feces and urination, find new water sources, and detour around the impacts of grassland fires.[9] Historically, herds of wild grazing herbivores did not linger long in one location. The herd grazed down edible plant species fairly evenly, which allowed the landscape time to recover from the beneficial disturbance of the herd, including the manure left behind. Additionally, rivets in the soil left from the imprint of their hooves helped to hold and protect grass seed and capture water.

Modern-day herd management that mimics the behavior of wild herbivores bunches animals together into a single herd and employs tight control over the timing, intensity, and frequency of their impact on the land. *Timing* refers not only to the season of the year but to how many days the cattle herd spends in a specific pasture or paddock. *Intensity* refers to the degree of management of the movement of animals through a pasture area. *Frequency* measures how long a paddock is rested from grazing before a herd returns, an interval of time that can

Contract Grazing by SweetGrass Grazing Service is implemented for fuel load reduction within wildland-urban interface landscapes in Northern California. *Photo by Aaron Gilliam.*

vary considerably depending on ecological objectives and conditions. There are various schools of thought on the specifics of implementing managed grazing, and some of the terms used to describe these practices are Holistic Planned Grazing and Adaptive Multi-Paddock (AMP) grazing, as well as short-duration, management-intensive, pulse, and cell grazing. The principles arise from the ideas of French farmer and biochemist André Voisin and later advanced by Allan Savory, who honed in on these techniques after years of observing wildlife behavior and ecosystem responses in southern Africa.

Rangelands need periodic pulses of disturbance, such as fire or grazing, in order to remove dead or moribund grasses so that new plants can grow and thrive when the rains come. Too much rest is as bad for the land as is too much grazing. Fire can easily cycle old and dead grass into nutrients for new plant life, but so can grazers, such as deer, antelope, bison, sheep, and cattle, which return nutrients to the soil in the form of urine and manure. The timing, intensity, and frequency of animal impact inherent in AMP grazing, often controlled by temporary electric fencing or a human herder, can be implemented to stimulate plant vigor and

Kaos Sheep Outfit is operated by Robert and Jaime Irwin. Their flocks graze in vineyards, orchards, and on private and public rangelands to reduce the fuel load in Northern and Central California.

resilience while allowing natural cycles, especially the carbon cycle, to operate as nature intended.

In contrast, continuous grazing—the standard management scheme on most of the world's ranches—happens when livestock are allowed to disperse uncontrolled across the same pastures day in and day out, year after year, eating their favorite plants down to the ground and ignoring less well-liked ones, which causes severe damage to the land. This type of grazing management can lead to biological stress on plants, soil erosion, and even outright desertification. Unfortunately, despite the obvious advantages of AMP grazing, there has been ongoing and at times heated debate among researchers, environmental activists, and some ranchers concerning the merits of continuous versus planned livestock management.[10] Recently the issues of climate change, carbon sequestration, and soil health have been injected into this debate.

One attempt to resolve this dispute has been led by Dr. Richard Teague, a rangeland ecologist at Texas A&M University, working collaboratively with colleagues from Wisconsin, Ohio, and Michigan. They compared the soil carbon impacts of continuous grazing versus Adaptive Multi-Paddock grazing on the Edwards Plateau Ranch, located in north-central Texas, and discovered that the AMP grazing test plots sequestered three tons (2.7 metric tons, equivalent to one-sixth of an individual's annual greenhouse gas contribution) *more* carbon per hectare than the continuously grazed systems while improving the plant species composition, decreasing the amount of bare ground, and enriching soil microbial composition. Additional research by Teague and his colleagues determined that the carbon storage in soils as a result of AMP grazing more than offset methane emissions from cattle. "Incorporating forages and ruminants into regeneratively managed agroecosystems," they wrote in a peer-reviewed paper, "can elevate soil organic C, improve soil ecological function by minimizing the damage of tillage and inorganic fertilizers and biocides, and enhance biodiversity and wildlife habitat."[11]

In the fibershed model there is a significant link between restorative rangeland management and fiber production, especially when we consider the natural behavior of sheep, goats, alpaca, and other fiber-producing herbivores. These animals originally moved around a landscape according to the presence or absence of predators and forage availability. Later they were shepherded by humans who guided them to new pastures as the seasons came and went. Traditionally this activity has been called *pastoralism*, and remnants continue into the modern era led by herders who have worked hard to retain their culture's pastoral heritage and practices (the cattle-herding Maasai of eastern Africa are a good example). In contrast, modern industrial agriculture, including the use of synthetic biocides and continuous grazing, strips the land of its health, resulting in the degradation of soils, long-term declines in plant productivity, decreased biological diversity, a negative impact on farmer income, and an overall reduction in the resilience of the system.[12]

While there are examples of overgrazed pastoralist systems, the question today is how to encourage the adoption of practices that regenerate soil while staying true to a region's cultural values—all while drawing down atmospheric carbon. One answer is simple: As wearers of clothing, we have the opportunity to protect and enhance historically rooted fiber systems that stimulate the land's natural self-renewal capacity. Our support for these systems as wearers entails identifying the shepherds, shepherdesses, ranchers, and farmers who are committed to restoring the land, and once they're identified it is our job to see their efforts proliferate by creating a demand for clothing made from their finest fibers and filling our pillows, mattresses, and woven carpets with their coarser fibers.

In the Soil-to-Soil concept, the role of the shepherds and other land managers involves continuous, contemplative stewardship focused on a

Herding Schools
by Brittany Cole Bush

How do we find and teach the next generation of shepherds? For decades Spain and France have met their need for sheepherders through the sponsorship of well-attended training programs that have graduated hundreds of pastoralists. Many applicants to these schools are young people from cities who desire to work in agriculture. Among the skills gained, students learn how to use voice commands and herding dogs to carefully control the placement of a flock of sheep to achieve nutritional and ecological goals.[13]

Spain has five herding schools, one of which is based in the Pyrenees Mountains in Catalonia, called Escola de pastors Catalunya. It is run by Associació Rurbans, a nonprofit organization dedicated to supporting and preserving the cultures and landscapes of the region. The school's curriculum focuses on providing a scientific foundation in animal husbandry and ecology combined with real-world herding practice and mentoring in the production of dairy, meat, and fiber. Over the course of the six-month training, the students, whose average age is thirty, develop their own projects and business plans, which allows to them graduate from the school with a road map for practical application of the knowledge and experience gained.

The Escola collaborates with the wool growers' association Xisqueta Obrador, which has created a wool brand known as Xisqueta, the name for the regional Merino sheep breed. The brand brings regionally produced fiber products to market from animals that have been tended by the school's shepherds. From the pastoral mountain landscapes where the sheep graze, valuable raw wool is produced, harvested, and processed into yarns that local artisans work with, often using natural dye techniques while collaborating with designers to create place-based clothing. Another example of collaboration is Serradet de Borneres, a small-scale sheep operation that is owned and operated by alumni of the herding school. It produces sheep cheese that is sold directly into local markets.

Achieving the goals of cultural preservation, supporting a regional food and fibershed, and employing environmentally sound management of their flocks, these businesses demonstrate how herding schools can train and empower new farmers and pastoralists as innovative entrepreneurs.

refined understanding of the conditions that restore and maintain the land's capacity to support life. It is essential to provide grasses and forbs well-timed rest periods from grazing as well as to make decisions about where and when to graze based on reading the signs of the entire ecosystem. By deepening our understanding of the historic ecological role of ruminants, we will be able to start a proactive conversation about how to improve and refine our engagement with large herbivores, including cattle. However, too often we become distracted by ideologically driven arguments that originate from behind desks or on social media platforms about what agriculture needs to look like—like who's eating what and how—which causes us to lose sight of larger ecological questions. What we could be asking is this: What is the proper role of ruminants on our planet at this time, and how can we enhance our relationship with them to repair and restore earth's ecology?

The Importance of Carbon Farming

California's most recent lengthy drought, which began in 2013 and persisted into 2017, was so serious that it arguably had no historical precedent and provoked a group of collaborating organizations in my fibershed to hasten the implementation of soil carbon-building projects throughout our region's working lands. The need for increasing our soil's ability to retain moisture was becoming a stark life-or-death reality. The conversation that emerged focused upon the ecological function of *whole* farms or ranches. In other words, if we focus on how to enhance the quantity of the sun's energy brought into our entire farming system through its plants and the carbon they generate, the entire ecosystem benefits. This way of engaging with the land through a carbon lens has generated a body of practices known as *carbon farming*. Some call it *ecologically enhancing agriculture*, but calling it carbon farming keeps our eyes on the goal of drawing down greenhouse gases by focusing on the principal element that is both a source of concern and a tremendous benefit to society.

Carbon farming involves implementing practices that are known to improve the rate at which CO_2 is removed from the atmosphere and transformed into plant material and soil organic matter. It is successful when soil carbon gains resulting from land management exceed soil carbon losses. Famed American writer and conservationist Aldo Leopold said, "Land health is the capacity for self-renewal in the soils, waters, plants, and animals that collectively comprise the land." Carbon farming is an approach for engaging and sustaining this ability for "self-renewal." Practices that damage or destroy the ability for self-renewal should be discontinued, including tillage, overgrazing, using diesel as an energy source, and applying fossil-fuel-based fertilizers, pesticides, and herbicides. Practices that encourage

> Practices that damage or destroy the ability for self-renewal should be discontinued, including tillage, overgrazing, using diesel as an energy source, and applying fossil-fuel-based fertilizers, pesticides, and herbicides.

The Fibershed Carbon Farm cohort meets along a freshwater creek at Freestone Ranch to study the flow of the water and the impacts of a series of good winter rains.

self-renewal should be encouraged, including keeping the soil covered yearlong with living plants, minimizing soil disturbance, increasing plant and animal diversity, and managing livestock according to planned grazing principles.

Many of the practices and principles embodied in carbon farming have demonstrated their value globally but remain underutilized by farmers and ranchers, raising the question of how to inspire their adoption more widely. Agriculturalists often work with very tight financial margins and have to make difficult decisions on a day-to-day basis that frequently involve constantly changing variables, including the weather, animal health, equipment function, and labor. These constraints impede the implementation of innovative practices, especially if they require cash or capital investments. To encourage a rancher to convert to carbon-enhancing management, a new approach to financial incentives is required, one that not only reduces the financial risk that comes with changing practices but also encourages farmers and ranchers to move quickly—for good reason. Christiana Figueres, executive secretary of the United Nations Framework Convention on Climate Change, has stated that in order to keep our planet from exceeding a crippling two-degree

Jacob sheep rotating to a new pasture at Robin Lynde's Meridian Jacobs farm.

Celsius rise in global temperature, emissions must peak by 2020. "Some say that is impossible but is impossible is an attitude, not a fact. Agriculture has a critical role to play, both in dramatically reducing emissions and by providing a sink to draw down carbon from the atmosphere," Figueres told a room of scientists, farmers, and policy makers during a 2017 Sequestering Carbon in Soil conference held in Paris, France.[14]

In the United States recent developments have allowed the capture and storage of atmospheric CO_2 in soils to become both practical and verifiable. One is the Carbon Farm Planning (CFP) process developed by Dr. Jeffrey Creque, a rangeland ecologist with the Carbon Cycle Institute in Petaluma, California. The CFP focuses on increasing the capacity of the working farm or ranch to capture carbon and store it in vegetation and soil organic matter. CFP is based upon the planning process developed by USDA's Natural Resources Conservation Service but uses carbon and carbon capture as the organizing principle around which agricultural plans are constructed, connecting on-farm practices directly with ecosystem processes.

Carbon Farm Planning starts with an inventory of natural resource conditions on the farm or ranch but with a focus on opportunities for

No-till planting directly into crop stubble from the prior growing season. *Photo by Maggilautaro.*

reduction of greenhouse gas emissions and enhanced carbon capture and storage. The inventory is transferred to one or more maps of the ranch, which are then used to mark potential carbon-capture locations such as silvopasture (incorporating trees in a pasture system) or stream restoration projects, with the overall goal of envisioning how the farm or ranch may be expected to perform years into the future. Next, the carbon benefits of each project and agricultural practice are quantified to estimate how many tons of carbon dioxide and other greenhouse gases would be either *reduced* as an emission or *sequestered* in the soil. A menu of practices and benefits is then developed and prioritized based on the needs and goals of the farm or ranch, and economic considerations are factored in, including potential funding sources. After implementation, the CFP is monitored and updated to meet changing conditions, objectives, and new opportunities while sticking to the overall carbon goal.

The modeling program that estimates and tests greenhouse gas emissions and soil carbon sequestration is called the Carbon Management and Evaluation Tool (COMET) and was developed by Colorado State University. While the tool was originally designed for commodity agriculture, the CSU team were open to refining it for a number of different farming practices and specialty crops. Since then the COMET model has grown to cover many diverse on-farm practices. Today it has become a platform for farmers and ranchers to estimate greenhouse gas emissions from their activities and assess scenarios to reduce those emissions and increase carbon in soils and vegetation. COMET allows agriculturalists to run "what-if" scenarios about how changes in their crop and livestock management could have a climate impact and then track the implemented changes on the landscape over time.

According to Dr. Keith Paustian (a lead scientist who works on the model), COMET is based on more than forty different greenhouse gas emission models developed by scientists. "Each of the models predicts the greenhouse gas emissions or carbon sequestration associated with a different aspect of farming crops and managing livestock," Paustian said, "such as applying fertilizer, growing trees, storing livestock manure, or managing organic matter in soils. The model was developed using several decades of research in soil carbon and nitrogen dynamics, and it has been validated against literally hundreds of experiments."[15]

Included in the first cohort of carbon farm plans were two ranches in drought-impacted areas in California—Modoc and Santa Barbara Counties. At the time, these ranches had been heavily impacted by multiple years of below-average precipitation. Jeffrey Creque, local resource conservation districts, and representatives from the Natural Resources Conservation Service worked in these locations to draft a management plan for the family-owned ranches that included multiple and layered strategies for improved carbon capture, including windbreaks, hedgerows, freshwater (riparian) creek restorations, managed grazing plans, and the addition of compost to the rangelands. In the Modoc County ranch, the greenhouse gas impact was measured in carbon dioxide equivalents, and the forty-five-hundred-acre land base was determined to have the ability to sequester 111,581 metric tons of CO_2e over the twenty-year span of time required to see many of the projects into maturity. This is equal to the emissions produced by 23,740 passenger vehicles in a year. The increase in soil water-holding capacity at the ranch was measured at over 520 acre-feet (169.5 million gallons or twenty-two million cubic feet). This amount of water-holding capacity is equivalent to the amount of water used by 1,730 California households in a year.[16]

In the case of the Santa Barbara carbon farm plan, which was developed for an eight-thousand-acre ranch, the estimated sequestration potential over a twenty-year period is two hundred thousand metric tons of CO_2e, the amount of emissions produced by 43,478 passenger vehicles

Sarah Keiser from the Penngrove Community Grazing Project inspects a young oak that has been protected from grazing pressure at Freestone Ranch near Sebastopol, California.

in one year. If implemented the plan will also increase the water-holding capacity of the soil by nine hundred acre-feet (293 million gallons or thirty-nine million cubic feet), equivalent to thirty-one hundred households—a savings that would exceed the ranch's current surface water capacity by 1,800 percent.

Predicting carbon and water impacts through the use of the computer models that exist today is currently the most affordable way to enable understanding of the impact of land management. There are systems emerging that will hopefully become more common and accessible in the future, focused upon collecting farmers' and ranchers' real-time data (through sensors and satellite imagery); this would provide immediate feedback on ecological impacts of management decisions. Pilot projects are being initiated to establish new economic platforms that will allow ecosystem service payments to be made to farmers and ranchers for creating ecological value based on real-time data derived from working landscapes. The work to monitor ecological function in present time is currently expensive to scale. Like all new technologies it will be important to continue to direct its usage for the purpose of closing wealth gaps and establishing greater value for the work that humans do that builds lasting prosperity for future generations.

The Promise of Agroforestry

One of the most important tools in the carbon farming tool kit is the planting and careful integration of trees and woody shrubs into crop and animal production systems. Called *agroforestry*, it has many benefits for a landowner. It can help to control pests by providing habitat for beneficial insects and birds; reduce cold and heat stress on animals by providing shade and shelter; slow runoff to reduce flooding, soil erosion, and water pollution while increasing infiltration; protect crops from wind damage; trap snow for additional soil moisture; and provide multiple food and commercial products, including fruit, nuts, timber, fence posts, and wildlife habitat.

Agroforestry comes in many forms. Here are a few:

RIPARIAN FOREST BUFFERS: Streamside plantings of trees and shrubs that reduce water pollution, stabilize bank erosion, and provide shade for aquatic environments.

FOREST FARMING: The cultivation of high-value non-timber crops (food, medicine, and crafts) under the protection of a forest canopy that has been managed to provide a favorable crop environment.

WINDBREAKS AND SHELTER BELTS: Rows of trees and shrubs that reduce wind speed. They improve crop yields, reduce soil erosion, improve water efficiency, protect livestock, and conserve energy. These plantings can increase carbon storage in biomass and soils, reduce soil erosion and loss of soil moisture from wind, protect pastures and crops from wind-related damage, improve the microclimate for plant growth, provide shelter for livestock, enhance wildlife habitat, provide noise and visual screens, and improve irrigation efficiency.

ALLEY CROPPING SYSTEMS: Widely spaced rows of high-value trees that create alleyways for crops. They benefit trees and crops, providing annual and long-term cash flow. Planting grain crops within these alleys and grazing that grain crop down early in the season and after the grain harvest adds carbon and nutrient cycling benefits to this system.

SILVOPASTURE SYSTEMS: A combination of trees with forage and livestock production on the same field. The trees are managed for wood while providing shade and shelter for livestock. Trees can provide long-term economic returns while livestock and forages generate an annual income from the same pasture. Correctly managed, the combined production can be greater than separate forestry and forage livestock systems. Adaptive Multi-Paddock–style livestock management is required, particularly in the early years during tree establishment. A good example of an integrated crop-and-livestock system can be found in Northern California, where some vineyards are grazed by sheep (discussed more in chapter 6).

To get a sense of how carbon farming could positively affect global warming, imagine ten ranches, approximately three-hundred acres each (land area needed to operate a medium-scale grazing operation in our region), implementing carbon farming and achieving a conservative per acreage carbon sequestration rate on an annual basis. At this scale, and

Community members and movement leaders model Climate Beneficial homegrown, woven, and sewn designs, many of which were constructed from Bare Ranch Wool woven by the Huston Textile Company.

with conservative carbon drawdown figures estimated, the total net gains for carbon and water would equate to 120,000 metric tons of CO_2e removed from the atmosphere over 20 years, equal to the emissions that 25,532 passenger vehicles produce in a year, and an increase in soil water-holding capacity of 660 acre-feet (215 million gallons or twenty-eight million cubic feet). That's 2,250 households a year. These numbers illuminate the tie between water and carbon and reinforce how we can support working lands and become increasingly more resilient to long-term dry periods by focusing our attention on carbon in the soils.

A Solution: Climate Beneficial Fiber

There is no one correct way to grow climate-friendly fiber. Instead, fiber-sheds can encourage a place-based approach that empowers farmers, ranchers, and other land managers to explore the wide variety of car-bon-capture opportunities. While the practices may differ from ranch to ranch, the overall goal is the same: the enhancement of atmospheric carbon capture. We can solve the climate crisis only through a paradigm shift in land management. A term that has become popular recently to describe this paradigm shift is *regenerative agriculture*—which can be defined as a system of agriculture that restores and maintains nature's capacity for self-renewal while increasing on-farm productivity. At Fibershed we use the trademarked term Climate Beneficial to describe and verify the materials we are growing on ranches and farms that are implementing carbon farm plans. Because our regional agricultural systems have lost considerable levels of carbon over the centuries (more than 50 percent in our rangelands alone), we have a carbon debt that we owe to our soils that we are actively making up for through carbon farming. Every metric ton of carbon we move from the atmosphere to the soil benefits our climate, and by virtue of this measurable action, we are confident to verify our efforts as Climate Beneficial. Replacing this carbon debt is the necessary first step in establishing agricultural systems that will eventually maintain their capacity for self-renewal, and therefore become truly regenerative.

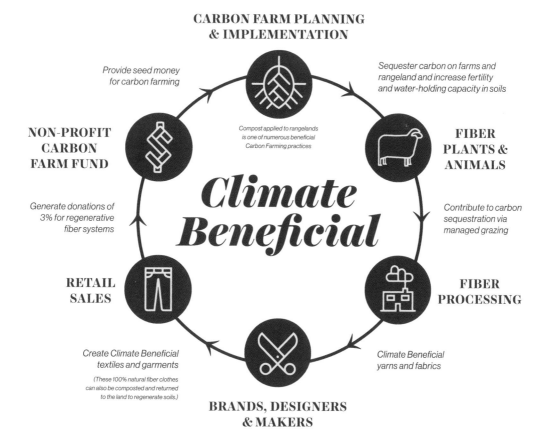

The elements of the Climate Beneficial process. *Illustration by Andrew Plotsky, courtesy of Fibershed.*

A critical planting area at Loren Poncia's ranch is implemented using a selection of native species selected by STRAW (Students and Teachers Restoring a Watershed). *Photo by STRAW.*

A significant number of sheep are grazed upon rangeland, pasture, and perennial and annual cropland systems. Alpaca, llama, and mohair producers also utilize regional pastures and to some degree rangeland for their agricultural practices. All of these grass-fed, fiber-producing animals have the potential to graze on managed landscapes where carbon farming practices are being implemented, thus we are able to create products that are labeled as Climate Beneficial by virtue of their integral place in the carbon farming system.

However, there are some challenges to moving more expensive Climate Beneficial goods into the marketplace. Consumers are generally averse to higher pricing, and there is a general lack of understanding about the role of agriculture as a climate solution (particularly animal agriculture). Acknowledging those concerns is important, but we need to move toward more in-depth conversations. For example, when you break down the purchase of a well-made but more expensive garment into "cost per wear"—and you purchase fewer garments overall—you quickly find that a well-made garment, especially a sweater or coat, will cost pennies per wear, and in most cases the garment will outlive the wearer. In the case of raising animals for fiber, this topic deserves another book altogether, but it is hopefully evident thus far that herbivores are critically important to the health of soil systems (both food and fiber) and are an important asset in the amelioration of the climate crisis.

Consumer demand for natural fibers and ethically, locally, and domestically produced goods is on the rise nationwide, along with a desire for increased supply-chain transparency and material traceability from textile manufacturers and retailers. It is logical to think that consumer demand could also create increased value for wool producers who provide domestically grown, traceable wool. One way to reassure consumers that Climate Beneficial fiber is worthy of the added cost is through a process called a *life cycle assessment* (LCA), which is used to track environmental footprints (water and carbon primarily). Within the textile industry, most carbon-focused LCAs assess the footprint of a garment specific to the emissions of greenhouse gases from mechanical production—"smokestacks and tailpipes"—as well as emissions that are broadly and globally estimated

> One way to reassure consumers that Climate Beneficial fiber is worthy of the added cost is through a process called a *life cycle assessment,* which is used to track environmental footprints.

based on the total land area that a fiber requires for its growth and production. However, from what we now understand specific to the soil carbon pool and the ability of this pool to be managed for carbon sinking (rather than carbon emitting), we have the tools to manage and measure carbon impact in a very different manner—starting with where the fiber is grown or raised. We now know that an LCA that takes into account only total land area and manufacturing emissions is an incomplete picture of a garment's contribution to global warming. Thanks to an emerging and increasingly refined place-based biogeochemical understanding, we are able to measure soil and atmospheric carbon exchanges on fiber-producing landscapes in our home community where carbon farming and other soil regenerating practices are occurring.

In 2013 Fibershed partnered with Dr. Marcia DeLonge of the University of California–Berkeley's Silver Lab during our Wool Mill Vision research project. DeLonge analyzed seven scenarios that represented a broad range of land management and production practices for a wool garment, scenarios that illuminate a range of carbon footprints with high to low greenhouse gas emissions. What was unique about this life cycle assessment is that it utilized localized peer-reviewed soil carbon sequestration research for grazed rangelands.[17] The LCA was site-specific and encompassed a set of soil types within our part of California, including soils from the coast to the Sierra Nevada, a region that produces over a million pounds of wool per year, one-third of what California produces per year in total.

DeLonge explored these seven scenarios for greenhouse gas impacts of various phases of the life cycle of a wool garment: the land management, animal productivity, animal emissions, transportation of raw and processed materials, employee commutes, manufacturing systems, and end-user care of the garment. The first scenario analyzed a higher emissions scheme, known as conventional; the five subsequent scenarios were various emissions improvements on this. All of the first six scenarios are net emitters of carbon dioxide. The seventh scenario represented a case where greenhouse gas emissions were reduced significantly through land management practices. The soil carbon sequestered within this scenario was modeled from measurements that were taken in what has become a decade-long analysis of the impacts to organic matter amendments on grassland soils.[18]

DeLonge's analysis demonstrated that if the rangelands could draw down three additional tons of CO_2e per hectare (concurrent with regionalizing the supply chain, adding renewable energy to the manufacturing systems, and localizing the workforce), it was possible to create wool

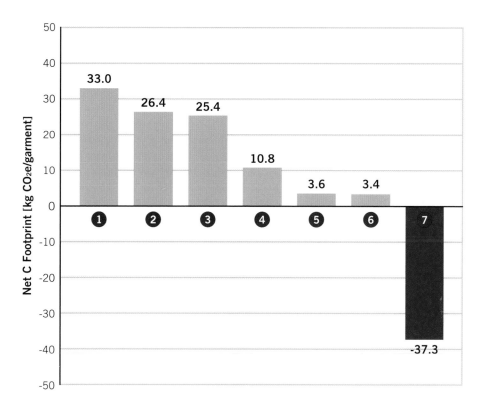

In an initial life cycle assessment (LCA) by Dr. Marcia DeLonge of UC Berkeley's Silver Lab, it was shown that sheep grazed on compost-applied rangelands can produce wool with a net carbon benefit. Subsequently moving this wool through a regional and renewable-energy-powered supply chain would produce a garment with a negative CO_2 footprint. *Graph by Fibershed staff.*

Conventional to Climate Beneficial Production for Wool Garments

(1) **Conventional Realistic:** CA grid-derived energy, slightly higher C footprint relative to other cases due to loss in soil C, synthetic fertilizer use, higher transportation costs

(2) **Conventional Optimistic:** CA grid-derived energy, but no increase in soil C

(3) **Fibershed Neutral Soil:** Geothermal-derived energy, but no increase in soil C

(4) **Fibershed Conservative:** Geothermal-derived energy, good land management increases soil C at a more conservative rate than case 7

(5) **Fibershed Realistic:** Geothermal-derived energy, conservative compost credit, good land management increases soil C at a more conservative rate than case 7

(6) **Fibershed Possible:** Solar-derived energy, conservative compost credit, good land management increases soil C at a more conservative rate than case 7

(7) **Fibershed Optimistic:** Solar-derived energy, optimistic compost credit, good land management increases soil C at optimistic rate, minor reductions in C footprint relative to other cases at several steps (transportation distances, commuter mpg, animal emissions, air-dried clothes, etc.)

Lani's Lana: A Ranch-to-Fiber Success Story

Lana means "wool" in Spanish, and Lani's Lana is the name of the fine-wool business owned by rancher and fiber artist Lani Estill. She works with her family to produce Climate Beneficial wool that is utilized by weavers, spinners, large-scale brands, and commercial yarn makers. Lani's wool comes from Rambouillet sheep that do well in the dry, open range of northeastern California; the wool they produce is perfect for use in garments. Additionally, some ewes are born with natural, non-white colors, such as gray and black, which makes their wool highly desirable as yarn.

What is, and has been, a traditional rangeland sheep operation has grown to include the work of making and applying compost, as well as developing shelter belts (pollinator corridors) and silvopastures. Lani and her family began working with Fibershed, the Carbon Cycle Institute, the Natural Resources Conservation Service, Point Blue, and a host of local contractors to produce Climate Beneficial wool beginning in 2016. Employing support from the NRCS and well-known outdoor company The North Face, the ranch began to implement the carbon farm plan that had been written by Dr. Jeffrey Creque in 2015. The plan includes a host of exciting projects, including riparian restoration (freshwater creek plantings) and miles of tree plantings to provide shade for animals, protection from wind, and above- and belowground carbon sinks.

Lani Estill's wool yarn and finished goods can be purchased from the Northern California Fibershed Marketplace, a project developed by the Northern California Fibershed Cooperative. The wool has been featured in numerous garments produced by regional designers, often in conjunction with the family-owned Huston Textile Mill, and featured in Fibershed's annual Gala show.

garments with a *net negative* carbon impact—meaning more carbon would be drawn down from the atmosphere than was emitted in the production and care of the clothing item. It was the soil's carbon drawdown capacity that provided the most significant positive impact. This LCA demonstrated that we could produce a wool garment with a positive climate change impact. In other words, sheep-grazed landscapes are capable of taking in significant levels of CO_2 and other greenhouse gases to offset (1) methane emissions from the animals (enteric fermentation); (2) the manufacturing systems; and (3) consumer use. Since this analysis was completed, our organization has galvanized its programmatic work around DeLonge's research, and we have sought to materialize a community-scale working model for the production of a Climate Beneficial garment.

The Soil-to-Soil framework addresses symptoms of the climate crisis by tackling root causes. This is done through employing cyclical

Lani's Lana Climate Beneficial wool yarns are dyed
in California-grown fermentation indigo and madder root.
The yarns are part of a shawl knitting kit designed by Kira K Designs.

The practice of adding a ½" layer of an organic amendment to California rangeland systems proved to support these soil systems in drawing down between one and three metric tons of carbon dioxide per year. *Photo by Marin Carbon Project.*

thinking and creating awareness of the transfer of carbon between carbon pools. This framework puts into immediate question any fiber or dye that damages our soils. Natural fiber and dye species farming can be integrated into existing agricultural models through integrated crop-and-livestock systems, the development of crop rotations, and the implementation of verified carbon farm practices as a means of rebuilding the 136 billion tons of lost carbon that we've released into our atmosphere over the decades. We've shown time and time again that farming can be a drawdown solution—and it *must* be, given the short window of action that catastrophic climate change has defined for us.

Further, the Soil-to-Soil system is grounded in the development of efficient and renewable energy-powered manufacturing systems that exist in proximity to where our fibers and dye plants are grown. End-of-life strategies for clothing are based on the returning of carbon to the soil from which it came (after many recycling and mending options have transpired). Natural fiber textiles that come from regenerating soils can and must be returned to the soil. This action both completes the cycle and initiates it all over again. Carbon from our clothing creates balanced compost when blended with nitrogen and mineral-rich inputs from manure and food waste.

It is important to remember that our planet's soils hold three times as much carbon as our atmosphere and four times as much carbon as the planet's vegetation. Soil represents the largest carbon sink that we have access to and in many cases directly manage via agriculture. We believe that understanding the power of soil carbon and the vital importance of living in balance with the carbon cycle is critical to our collective future. We also see that the most strategic redesign of our textile system is to focus on the development of regional fiber systems, for it is at the local level, where we carry the wisdom and understanding of our own land and watersheds, where we have the political power, the community connections, and the direct hands-on capabilities to make meaningful and significant changes to our businesses, school curricula, policies, consumption practices, and atmospheric concentrations of carbon dioxide. Regional-level management of resources and economies is where we've experienced the most traction for developing soil-based fiber systems solutions.

However, as exciting, hopeful, and scalable as these developments are for addressing climate change and encouraging local economies, recent discouraging developments in the textile world involving an extreme version of genetically modified proprietary organisms pose a substantial threat to progress. And so, before I can explore the fibershed model further in this book, I need to explain how this new threat—called synthetic biology—is a false promise.

A common feedstock for synthetic biology, sugar is often grown on plantations, such as this one in Brazil, that encroach on native rainforest. *Photo by BrazilPhotos / Alamy Stock Photo.*

CHAPTER FOUR

The False Solution of Synthetic Biology

*I*n the last decade venture-capital-backed biotechnology corporations have launched an aggressive marketing campaign to promote
their purported "scientific" solutions to the many global food and
fiber challenges spawned by the industrial production model. Abetted by
journalists and research funding from government agencies, these private corporations promote technologies they say can "redesign nature"
through cutting and pasting bits of DNA in order to "improve" everything from product shelf life to photosynthesis for the benefit of all. But
beyond the marketing spin, the primary goal of these companies is to
patent organisms that can then be sold to a marketplace of consumers
who unwittingly contribute to further concentration of power and profit
among a handful of investors and CEOs. Recently this moneymaking
scheme has expanded to include subjecting microbes to extreme DNA
engineering with the objective of creating "biological factories" that produce high-value, designer organisms and compounds that are supposed
to find their way to clothing racks and grocery store aisles (but very few
have so far). This extreme manipulation of DNA is called *synthetic biology*, defined by the ETC Group as a combination of engineering and the
life sciences brought together "in order to design and construct new
biological parts, devices, and systems that do not currently exist in the
natural world or to tweak the designs of existing biological systems."[1]

In an effort to tap into the $1.3 trillion global textile market, syn-
thetic biology companies are putting natural fibers and dyes in their
crosshairs. One fiber of focus is so-called spider silk, a material generated
from feeding genetically engineered yeasts industrially farmed sugars.
The public relations campaigns for these artificial fibers tout them as an

ecologically thoughtful next-generation green technology. The claims, however, do not include life cycle assessments grounded in the land-use impacts of the feedstocks that these synthetic biology systems require. Sugar, for example, is a common feedstock often sourced from corn, sugar beets, or sugarcane grown in monocultures employing the traditional suite of herbicides, pesticides, and artificial fertilizers. Sugarcane fields are regularly burned post-harvest, creating large amounts of airborne pollution while generating vast amounts of erosion and other types of land degradation. According to a study conducted on sugarcane fields in southern Brazil, 531 pounds of carbon dioxide equivalent were released to the atmosphere per ton of sugar produced. The major part of the total emissions (44 percent) resulted from residues burning; about 20 percent resulted from the use of synthetic fertilizers and about 18 percent from fossil fuel combustion.[2] "Sugar has arguably had as great an impact on the environment as any other agricultural commodity," wrote the World Wildlife Fund in a report. "Wholesale conversion of habitat on tropical islands and in coastal areas led to significant environmental damage, particularly a loss of biodiversity."[3]

Base life-forms (bacteria and yeast) are the building blocks for all other life-forms on our planet. The risks involved in reengineering them could prove to be more harmful to human and planetary health than any other genetically modified organism (GMO) currently on the market. The unfortunate part is that we won't actually be able to determine the outcomes of these biotech projects until it is perhaps too late. You would think that the potential for genetic pollution to create cascading and irrevocable impacts is reason enough for industry, regulators, and consumers to stop and think. Instead the alluring pitch has become omnipresent among the proponents of synthetic biology and is part of a broader effort by the industry to convince consumers, designers, clothing companies, and the media that synthetic biology is actually *the solution* to the vast offenses associated with industrial textile production, including the failings of Fast Fashion. This communications spin extends to industrial plant and animal agriculture—*Don't like the way your food, meat, or clothing is produced? We'll grow it in a laboratory instead*, say the biotech companies.

Promoters of synthetic biology are driven by the belief that biology is fundamentally programmable, and thus subject to engineering, just like any human-created technology. In other words, they believe life can be reduced to standardized components that can then be "recoded" into new configurations at the cellular level, much like a software program. The focus of synthetic biology experiments has been largely confined to

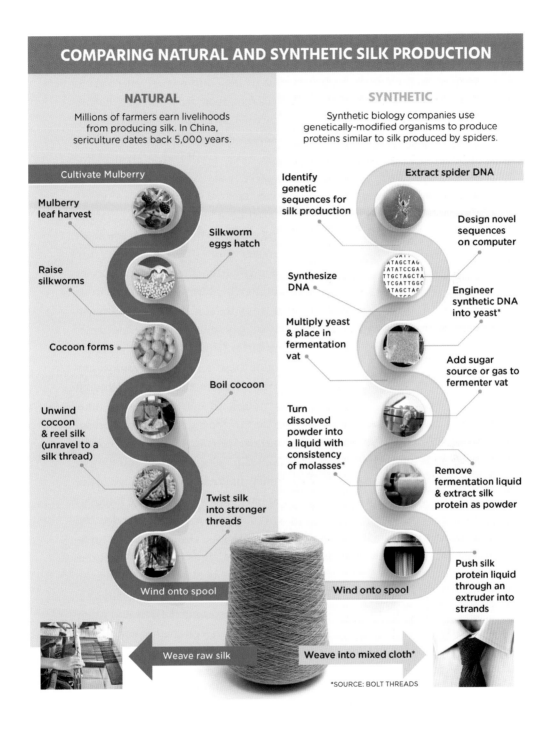

COMPARING NATURAL AND SYNTHETIC SILK PRODUCTION

NATURAL

Millions of farmers earn livelihoods from producing silk. In China, sericulture dates back 5,000 years.

SYNTHETIC

Synthetic biology companies use genetically-modified organisms to produce proteins similar to silk produced by spiders.

Cultivate Mulberry

Mulberry leaf harvest

Silkworm eggs hatch

Raise silkworms

Cocoon forms

Boil cocoon

Unwind cocoon & reel silk (unravel to a silk thread)

Twist silk into stronger threads

Wind onto spool

Weave raw silk

Identify genetic sequences for silk production

Extract spider DNA

Design novel sequences on computer

Synthesize DNA

Engineer synthetic DNA into yeast*

Multiply yeast & place in fermentation vat

Add sugar source or gas to fermenter vat

Turn dissolved powder into a liquid with consistency of molasses*

Remove fermentation liquid & extract silk protein as powder

Push silk protein liquid through an extruder into strands

Wind onto spool

Weave into mixed cloth*

*SOURCE: BOLT THREADS

"Comparing Natural and Synthetic Silk Production." *From "Genetically Engineered Clothes: Synthetic Biology's New Spin on Fast Fashion" by ETC Group and Fibershed. Diagram by Cheri Johnson.*

microbes, the workhorses of nature (*E. coli* is a favorite of engineers, for example). Microbes exist everywhere on the planet and perform a wide variety of critical functions, including helping plants to grow, digesting your food, producing oxygen, controlling pollution in the ocean, and regulating the underground universe of life. "Improving" microbes to create previously nonexistent types of fuel, food, and fiber has been a main target of research projects and financial investment firms in recent years, though scientists are also pursuing ornamental uses for synthetic biology, such as the creation of blue roses, and lifestyle changes for humans, like the elimination of foot odor.

Synthetic biology follows the path created by previous corporate efforts at producing GMOs over the past decades. Born in a laboratory, GMOs are the result of an artificial process by which one or more short sequences of DNA, called *genes*, are removed from the cells of a species and forcibly transferred into the cells of an unrelated species, creating a previously nonexistent "transgenic" organism. The source and the recipient of these genes can be almost any living creature on Earth, including microbes, plants, insects, and animals. In a lab genes can be recombined by bioengineers into artificial organisms with increasing speed, and thus the set of unknown consequences is allowed to accrue at a rapid clip. "Current understanding of the way in which DNA works is extremely limited, and any change to the DNA of an organism at any point can have side effects that are impossible to predict or control," wrote the authors of the website for the nonprofit Institute for Responsible Technology, a GMO watchdog organization. "The new gene could, for example, alter chemical reactions within the cell or disturb cell functions. This could lead to instability, the creation of new toxins or allergens, and changes in nutritional value."[4]

Cancer is one important side effect. Recombinant DNA and other forms of gene editing employ an enzyme (called a restriction enzyme) to cut the gene at a specific spot on the DNA strand, allowing a foreign gene to be attached. Recently a sequence of short, repeating DNA located within bacteria (where they fight viruses) called CRISPR has been determined by technicians to be useful for the production of basic biotechnology products, particularly when edited by an enzyme called CRISPR-Cas9. Once edited, a gene proceeds on its usual biological trajectory (except for the natural selection part) to become part of a microbe, plant, or animal—*except* it is now a unique creation. In this uniqueness lies the trouble. Once released, these new organisms "cannot be recalled, and any changes to the genetic makeup of the population they induce are most likely irreversible," wrote scientists Janet Cotter and Dana Perls in a report for Friends of the Earth. "Hence, the genetic changes to a population are likely to persist for a very long time, possibly permanently. This may result in far-reaching and unpredictable consequences for society and the environment."[5]

Two recent studies determined that genes edited by CRISPR-Cas9 with the intention of treating disease increased the risk of triggering cancer in host cells, sowing tumors in patients.[6] Cells, researchers discovered, when injured by CRISPR-Cas9 in the editing process, either mend the DNA break or self-destruct, increasing the cancer risk. In another study scientists in the UK determined that CRISPR may cause "genetic havoc" by its disruptions.[7] These discoveries raise a huge amount of uncertainty about CRISPR—and gene editing generally.

GMO Disasters

Unfortunately we've been down this dangerous high-tech, profit-driven road before in the textile system. In the early 1990s Bt cotton was commercialized by biotechnology giant Monsanto. (Bt, short for *Bacillus thuringiensis*, is a toxic bacteria that kills moths, butterflies, beetles, and cotton bollworms.) The bioengineered cotton produces the Bt bacteria as a function of the plant's growth so that all insects die that try to feast on it. But there's more: In 1997 Monsanto commercialized Roundup Ready Cotton—a genetically engineered cultivar that is largely resistant to the use of the herbicide known as Roundup, which is sold by Monsanto as a companion chemical alongside the engineered seeds. The active ingredient in Roundup, *glyphosate*, has been banned in parts of the EU and has been recently labeled as a probable carcinogen in California. Not long after commercialization, Roundup Ready Cotton made its way to Indian agricultural systems through Monsanto's Indian partner Mahyco Biotech. In 2001 in Gujarat, India, large tracts of land were planted with Bt cotton varieties without prior government approval. Local officials subsequently discovered this had been going on for years without any government oversight, and when local farmers had their open-pollinated, classically bred cotton seeds tested, it was discovered that the Bt gene had already contaminated them. The farmers demanded legal action, and as a result the Bt crops were uprooted and burned by authorities.

Bt cotton was approved for use in the United States in 1994. Bt stands for *Bacillus thuringiensis*—a bacterium toxic to the larvae of moths, butterflies, beetles, and cotton bollworms that has been inserted into cotton in a manner that causes the plant to produce the insecticide in its tissues. *Photo by Kimberly Vardeman.*

Soon afterward, however, Mahyco Biotech was provided legal authority by the Indian government to sell the genetically engineered cotton in six Indian states. The decision was based on a flimsy argument that there was no point in denying approval since so many illegal varieties of Bt cotton were already present in the fields, and contamination of the seed stocks had already occurred. ("Contamination," in this case, means the genetically modified crops have cross-pollinated with the naturally, classically bred crops, altering the genetic heritage of the region's seed.) But when Bt cotton was formally introduced to Andhra Pradesh, Haryana, Karnataka, and Maharashtra in the early 2000s, shepherds began to watch their cattle fall ill after grazing on the cotton crop stubble, in some cases resulting in death. As a result, in four of India's cotton-growing districts in Andhra Pradesh, hundreds of farmers filed formal complaints with their joint directors of agriculture requesting compensation for crop losses due to yields far below what the company representatives had promised. The biotech companies appealed, and finally the appellate committee upheld the awards for the farmers.

The impacts of the genetically engineered cotton have negatively woven their way into the financial lives of Indian farmers over the last twenty years. Bt seeds and their accompanying chemicals must be purchased every year; these are not seeds that a farmer can save, and so by growing Bt cotton a farmer becomes a dependent on the company that is selling the seed and the accompanying chemicals. And because throughout India pests and weeds have become resistant to the herbicide and the Bt gene, farmers have been in turn forced to purchase much larger quantities of herbicides and pesticides to try to avoid crop loss. The cycle of debt for the world's poorest farmers has hastened with the entry of pricy seed and chemical inputs. The debt has led to such destitution that by 2013 over 270,000 farmers had committed suicide over concerns of money and poor crop yields.[8] "Industrial food and industrial fashion have destroyed lives and livelihoods and polluted soils and oceans," said Indian food and farm activist Vandana Shiva. "Over two decades, I have witnessed and studied the crisis for Indian farmers triggered by genetically engineered Bt cotton, which was introduced on the promise of pest control and reduction of pesticides but has created an epidemic of pest attacks, pesticide deaths, and farmers suicides."[9]

This example highlights a narrative promulgated by biotech companies that yields of genetically engineered crops are high enough to justify the cost of seed and chemicals. As each farming season unfolds, however, the holes in this narrative become larger and are now very hard to miss. By 2016, in Burkina Faso, one of the world's top cotton-producing nations, farmers had fully abandoned Bt cotton due to poor-quality fiber and shortened staple length. What is so critical in assessing these bio-tech projects is that fundamentally not one cotton farmer in India (or any other country for that matter) requested a seed with a built-in toxin or resistance to Roundup. However, many of these farmers have asked for the value of their labor to be better represented in the price of the raw material, and communities across India—for more than a century—have asked for the right to process their own cotton and be able to sell their finished textiles in a manner that benefits their community-scale economies. The *khadi* movement spearheaded by Gandhi remains an integral part of the social and cultural aesthetic of India today. The movement was and is based in a vision for economic and cultural sovereignty for the Indian people, rooted in the actions of artisanal processing of Indian fibers with simple machines like the charkha spinning frame.

As for Roundup itself, in August 2018 a jury in San Francisco awarded $289 million in compensatory and punitive damages to groundskeeper Dewayne Johnson for the suffering he has endured as a result of his cancer, which he claimed was caused by repeated exposure to the herbicide. The jury agreed with Johnson's lawyers that Monsanto was liable and that scientific studies back up their argument. For example, a study published the journal *Toxicology* showed that glyphosate disrupted human cells within twenty-four hours of exposure in sub-agricultural doses.[10] DNA damage occurred with exposures as low as 5 ppm (parts per million), and endocrine disruption occurred at as low as 0.5 ppm. In another study epidemiologists connected endocrine disruption to metabolic syndrome, obesity, and type 2 diabetes.[11] Residues of glyphosate on some livestock feed are allowed at levels up to four hundred ppm, and there are currently no laws regulating the concentration of glyphosate residues on our fibers. There are also no current laws regulating the concentrations of glyphosate in our surface water, rainfall, or soils. In our cotton-growing belt in California, the US Geological Survey found that 65 percent of lake, river, and stream samples in the San Joaquin–Tulare Basin had detectable levels of glyphosate.[12]

One more example of the perils of GMOs: In 2001 ecological geneticist Ignacio Chapela, a professor at UC Berkeley, and David Quist, a graduate student, published a startling paper in *Nature* announcing their discovery that many heritage varieties (called landraces) of maize in Oaxaca Valley of central Mexico had been contaminated by pollen from GMO corn, including farms found high in the mountains above the valley floor.[13] The GMO corn, which originated in the United States, was planted illegally in defiance of Mexico's 1998 ban on the product. Chapela and Quist's

Classical Cotton Breeding

Should we be enhancing nature's time-tested processes or investing in the false promises of bioengineering? The choice is a clear line in the sand that divides millennia of human efforts. Fortunately there are seed breeders who continue building upon the work of our collective plant breeding ancestry. In the fiber-farming world, there is one woman, Sally Fox, who continues to enhance the beneficial traits of the two major cotton species, *Gossypium hirsutum* and *G. barbadense*. Located in the Capay Valley in Northern California, Fox has been working for thirty years to develop her own certified organic varieties that are now known and loved across the world.

Any plant breeder can receive seed from the USDA seed bank. And any plant breeder, regardless of scale, can grow the plants, collect seed from them, develop new traits through classical breeding practices, stabilize those traits, name their variety, and fill out the paperwork for plant variety protection (PVP). This system of sharing advances the collective genetic pool while giving credit where it is due. Unlike the more recent patenting processes of large genetic engineering seed companies, PVP allows for generation after generation of seed breeders to expand and enhance upon one another's work. "There is a huge distinction between PVP and patenting," said Fox. "As a breeder you are participating in

the advancement of everyone else; patenting a trait halts that collective advancement potential. You cannot save patented seed or give it away."

Fox has been breeding cotton plants for traits such as color (green and brown fiber are the most famed) but also for traits such as staple length, color- and lightfastness (how the fiber performs in a textile), and yield. Her work in the field consists of perceptive observation followed by medically precise actions. A season's work will include cross-pollinating plants to produce what are known as F1 (first generation) seeds. The first-generation plants are grown the following season and are 100 percent hybrids. The seeds they produce are known as F2 (second generation) seeds. From these F2s, which are planted in the third season, Fox begins to select seed. She then continues to select for many generations, season after season, until the seed is stable and 100 percent consistent with the parent plant.

The complexity and beauty of these cotton varieties runs deep, and Fox's knowledge covers every facet, from the science to the cultural history. She is a teacher by nature, and she is now taking the time to cultivate a curriculum passing along her breeding knowledge to the next generation. To learn more about Fox's work and nonprofit development, you can begin by visiting her farm website: www.vreseis.com.

conclusions were attacked by pro-GMO advocates, including fellow researchers at Berkeley (the university at the time had received a large research grant from the biotech giant Novartis), forcing the editors of *Nature* to issue a partial apology. However, a subsequent study in 2009 confirmed Chapela and Quist's original conclusions about landrace maize contamination in the heartland of Mexico.[14] The threat posed by GMO corn to a staple of Mexican diet, intimately linked to Mexican national

Tribal farmers of the Shiv Krishi Utthan Sanstha cooperative produce fair trade and organic cotton, harvested by hand. *Photo by Joerg Boethling / Alamy Stock Photo.*

identity, demonstrates how culturally and ecologically insidious genetically modified organisms are—and why they should be resisted at all levels.

Despite much evidence and data to show otherwise, biotechnology companies maintain that genetically engineering life-forms will provide us the ability to contort pieces of the natural world into new shapes and functions that will meet all the needs of humanity, while providing sizable profits to those who patent the keys to these life-forms and the venture capitalists who back them. But what about the centuries of genetic heritage, the age-old seed swaps that provided free seeds to farmers and the ability to grow a crop without an accompanying chemical recommendation? All of that is lost when biotech systems come to dominate and reduce the genetic pool through their proprietary DNA.

As a methodology for interacting with DNA, the GMO approach does not reflect or mimic the natural world. Classical plant and animal breeding by humans, on the other hand, is a biomimetic process, meaning the selection is done in a manner that mimics a bee, herbivore, or carnivore. Nature does not reorganize the genetic building blocks of one species all at once, leaving the other interacting species within the ecosystem to their own devices. Would it not seem obvious that once you engineer cotton to become resistant to the boll weevil, the weevil would generate an equally powerful response; or if you're creating a cotton crop resistant to Roundup, the overhead spraying would eventually create resistance within another species?

Advances in genetics have a role to play in creating agricultural systems that can be regenerative, but it must be a careful one. "Molecular markers are being used now to speed up breeding," said Wes Jackson, founder of the Land Institute and a leading advocate for classical breeding techniques, "but much of what is being proposed by molecular biologists is greatly overstated, since plants still have to live outside, so ordinary breeding will continue to be necessary and, therefore, remain something of an art. Looking at molecules is to look downward in the hierarchy of structure. Often overlooked, but of equal importance, is the necessity to look upward in the hierarchy of the sciences to the ecosystem level, which makes ecology/evolutionary biology of greater importance to agriculture."[15]

Pandora's Box Gets Bigger

Recent developments in synthetic biology threaten to amplify the known risks posed by GMOs by orders of magnitude. Gene-editing tools are allowing start-up companies and researchers to alter DNA sequences in organisms more quickly and with greater precision (and greater risk) due to the use of programmed cell nuclei that allow the biotechnologists to manipulate the expression of traits and new functions anywhere along the DNA sequence. The process includes exploiting the natural repair mechanisms within the DNA, whereby the engineer will inject *donor DNA* into a cell; through the self-replicating repair process, this donor DNA becomes part of the target organism. Previously, recombinant DNA efforts, such as those used to make Bt cotton, inserted foreign DNA into living organisms at much smaller and narrower scales. Synthetic biology focuses upon recoding a cell's DNA, rather than recombining DNA from more than one organism in a traditional GMO process. The major risk is that the definition of *genetically modified*

> The major risk is that the definition of *genetically modified* could change enough to render the traditional warning labels attached to GMO products as insufficient.

could change enough to render the traditional warning labels attached to GMO products as insufficient. It also means the range of products that can be created via synthetic biology is huge and getting bigger, as are the uncertainties.

In his book *Regenesis*, which cheerleads the purported benefits of this engineering program, George Church, a professor of genetics at Harvard and owner of multiple patents, described synthetic biology as "the science of selectively altering the genes of organisms to make them do things that they wouldn't do in their original, natural, untouched state."[16] At the microbial level, he wrote, nature is being redesigned in increasingly ambitious and radical ways as engineers become adept at reprogramming living organisms. This is a good thing, Church insisted. "These technologies have the power to improve human and animal health, extend our life span, increase our intelligence, and enhance our memory." Why wouldn't we want to use synthetic biology to eliminate pathological viruses? Or attack cancer cells? Or bring back extinct species? Wouldn't it be great to see a woolly mammoth again? It's not an idle dream, Church declared. Not only will biology-as-technology enable us to re-create parts of our evolutionary past, we will be able to take evolution to places it has never gone before or would never go if left to its own devices.

In Church's worldview a living organism is like a computer, a ready-made, prefabricated production system that is governed by a program—its genome. Recoded, an organism could be made to produce practically anything, including our clothing, and then sold to us for a profit. And as for unintended consequences of this cellular reprogramming (called "off-target" effects in industry parlance), Church downplayed their significance, saying the benefits far outweigh the risks. However, this position is contradicted by a growing number of studies, including research that recently led the European Court of Justice to rule that synthetic biology procedures, such as CRISPR, must now be included in standard GMO regulatory evaluations.[17]

In an example of boosterism by advocates of synthetic biology, biotech companies claim they are now working on a "solution" to an important aspect of the industrial agriculture system—the synthetic nitrogen leaching crisis. Their purported goal is to reduce the amount of fertilizer applied to industrially grown crops, such as corn and wheat, but their real objective is to sell farmers a bioengineered product they don't need. Nitrogen is essential to plant growth and vigor. Plants receive nitrogen naturally and freely from the atmosphere (78 percent of air is nitrogen). This abundant overhead resource is drawn into the soil by plant species that have been around for millennia and that have co-evolved to cooperate with nitrogen-fixing microbes, which generate

Nitrogen is essential to plant growth and vigor. The biotech industry is designing custom-made microbes that fix nitrogen biologically for a profit, even though various plant species, such as this cover crop of fava beans, draw nitrogen into the soil naturally. *Photo by Nigel Cattlin / Alamy Stock Photo.*

Brown's Ranch and the Importance of Good Soil

In 1991 Gabe Brown and his wife, Shelly, purchased a farm near Bismarck, North Dakota, and began growing grains and raising beef cattle with heavy tillage and plenty of herbicides, insecticides, and artificial fertilizer. Three years later they stretched the conventional farming paradigm a bit by deciding to go no-till in order to conserve soil moisture and fuel costs. However, four successive years of weather-related crop failures created a desperate financial situation that set the Browns on a revolutionary journey from chemical, industrial agriculture to biological farming. Today Brown's Ranch is recognized as a leader in the regenerative movement for its successful integration of zero-till, cover cropping, soil health, mob grazing, and other practices that create ecological and economic health without relying on pesticides, insecticides, or synthetic fertilizer.

What many conventional farmers and ranchers view as challenges, such as soil compaction, erosion, weed pressure, and low yields, Gabe Brown views as opportunities because he considers them to be symptoms of problems, not causes. In fact, they are signs of poor soil health. His system, which imitates nature, focuses on regenerating resources by continuously enhancing the living biology in the soil. The results on his farm speak for themselves:

- Organic matter levels have risen from less than 2 percent to more than 6 percent.
- The biota in the soil has increased to the point that no synthetic fertilizer is now used.
- No fungicides or insecticides have been used on the ranch for over twelve years.
- Herbicide use has been cut by over 90 percent.
- Water infiltration and water-holding capacity are at their highest levels.
- A one-year drought is not worrisome— a healthy ecosystem adapts.
- Grain yields average 20 percent higher than the county average.
- Wildlife populations and species diversity have increased exponentially.

When Brown switched to no-till in 1994, he was the only one in Burleigh County, where 60 percent of the land is in grain production. Today about 70 percent of the county farmland is no-till. Brown thinks conventional farmers are beginning to understand, as he did, that their soils had become nothing more than a medium to hold the plants upright and that the dirt lacked not just soil structure and organic matter but *life* itself.[19]

Diverse perennial pastures on the Brown Ranch. *Photo by Gabe Brown.*

and deposit plant-available forms of the element into the soil. But in monocropped industrial systems, crops have become increasingly dependent upon ammonium nitrate fertilizer, chemically produced from methane, a potent greenhouse gas.

The biotech industry is now positioning their work as a solution to this problem: custom-made microbes that fix nitrogen biologically for a profit. That's the goal of Ginkgo Bioworks, a biotech firm working in partnership with the German chemical behemoth Bayer, as well as another company called Pivot Bio, a synthetic biology start-up based in California (and backed by a huge venture capital fund created by Bill Gates, Jeff Bezos, and other billionaires). "What everybody in the field would like to see with microbes is a renewable and sustainable way of producing that fertilizer," said Karsten Temme, CEO of Pivot Bio. "It really has been a long-term, elusive goal for the field."[18] Except it isn't elusive. Organic farmers know how to get nitrogen to the roots of their plants naturally with cover crops, polycultures, and the integration of livestock. They don't need custom-designed, artificially engineered, mass-produced microbes to correct the "inefficiencies" of nature—they just use more nature.

One way to ensure proper caution in the synthetic biology industry is through government regulation. However, a recent analysis concluded that there are troubling gaps in regulatory oversight of the fast-moving world of synthetic biology, most of them associated with the risk of bio-error—the unanticipated consequences of a mistake in the laboratory. The challenge arises from the nature of the organisms themselves, whose novelty, ability to evolve, complexity, and variability all present significant barriers to risk assessment by federal overseers. "It is not surprising that a technology as potentially revolutionary as synthetic biology would raise a number of concerns under a regulatory system developed largely prior to its inception," wrote the authors of the analysis. Effective regulation will require a major overhaul of the federal regulatory process at the EPA and FDA, and quickly, too—a job that the authors called "an extremely daunting task."[20]

This bleak assessment, supported by a report from Friends of the Earth, concluded that the EPA requires virtually no assessment of the environmental impact of these organisms, and the USDA gets involved only if pests are involved. For its part, the FDA has no mandatory requirement for food safety assessment. "Given the prevalence of unintended consequences from genetic engineering applications," wrote Janet Cotter and Dana Perls, the report's authors, "all genetic engineering techniques should fall within the scope of government regulatory oversight of genetic engineering and GMOs."[21]

The alternative is self-regulation—or what some synthetic biologists call "trust and then verify," which has become a mantra in the industry. However, history has clearly demonstrated how ineffective self-policing is among companies engaged in developing and promoting (and profiting from) a new technology. Meanwhile, the National Institute of Standards and Technology (NIST) launched the Synthetic Biology Standards Consortium in 2015 in order to standardize the design and documentation of synthetic biology across all academic institutions. This effort has hastened new research and encouraged start-up business developments. Unfortunately, even with considerable institutional backing from capital investors, taxpayers, and universities, the dangers and weaknesses of these new technologies are still not being fully discussed.

> Unfortunately, even with considerable institutional backing from capital investors, taxpayers, and universities, the dangers and weaknesses of these new technologies are still not being fully discussed.

It is not a surprise to learn that engineered biostructures can malfunction and cause an unwanted outcome—one that cannot necessarily be foreseen in computer models. Industry insiders agree. "The [synthetic biology] community is very much operating in a world where we cannot predict what is going to happen in our systems when we build them," said Reshma Shetty, cofounder of Ginkgo Bioworks.[22]

The uncertainties and risks are not stopping researchers or corporations from rushing into unknown territory, however. In a push for automation in order to expand scales of efficiency and productivity, mass-generation facilities called biofoundries are being set up around the globe, in which robotic assembly lines create and test new genetically engineered microbes. In turn these biofoundries are enabling researchers to embark on ever-more-ambitious projects. The assembly-line conditions can be harsh on microbes, which is why a search is on for different host organisms, including cell-free systems. The intellectual and financial returns could be huge. "I'm in this space because the frontiers are endless for what biology can do," said Shetty in the same article. "It's just a matter of the technology advancing to a point where those new horizons open up."

Investors and governments that support these biotech projects are looking only at the bottom line from a patentability and intellectual property standpoint and are not considering the aftermath for human health or their cultural impacts. As an example of the cheerleading for synthetic biology, the 2012 *Bioeconomy Blueprint* published by the Obama adminis-tration stated that "economic activity that is fueled by research and innovation in the biological sciences, the 'bioeconomy,' is a large and rapidly growing segment of the world economy that provides substantial public benefits. . . . It can allow Americans to live longer, healthier lives, reduce our dependence on oil, address key environmental challenges, transform man-ufacturing processes, and increase the productivity and scope of the agricultural sector while growing new jobs and industries."[23] The sales pitch sounds appealing, but what is never mathematically assessed within these economic blueprints are the post-innovation impacts fifty to one hundred years into the future from a cultural and biosphere-based perspective.

Meanwhile, the "gold rush" of synthetic biology is on. And just like the early days of the Industrial Revolution two hundred years ago, one of the first targets of this new technology is clothing, capturing the atten-tion of significant venture capital as well as the imagination of young fashion designers in schools such as Central Saint Martins in London. Hip young biotechnicians have spoken at events such SXSW, a popular media festival, and their work has frequented the TED Talk scene.

The potential new scales of pollution and non-composting waste are both arenas for evaluation that we need to be analyzing in advance of blindly accepting the material streams produced by synthetic biology. The world's oceans are already carrying microfiber pollution as a result of textile manufacturing, and the risks from machine-washing plastic-based clothing have become a major concern in recent years. The abra-sion caused during the mechanical washing process, after which effluent from our machines moves directly into water treatment plants that are unable to filter the plastic, has been noted as a source of microfiber pol-lution. Proponents of genetically engineered clothing are cleverly pitching their products as non-polluting. But the long-term degradabil-ity of bioengineered textiles cannot simply be assumed, according to the report by the ETC Group: "The proteins being developed by biotechnol-ogy companies are novel and distinct from natural spider silk and should be independently assessed for degradation claims and impacts once released into water and soil."[24]

There is also the question of what happens with the fermentation stock used to create these fibers and dyes and how this material impacts marine ecosystems—this issue has not been evaluated. We also do not yet understand the impacts to terrestrial or marine ecosystems due to accidental spillage of the genetically modified organisms that are part of the synthetic biology supply chain. According to the ETC Group, in 2010 a presidential commission acknowledged that "contamination by accidental or intentional release of organisms developed with synthetic biology is among the principal anticipated risks."[25] Yeasts in particular can spread quickly through the environment. Without proper regulatory and reporting safeguards in place, the risk of an accidental release of genetically modified microbes needs to be taken seriously.

Then there is the issue of taxpayer subsidy. For example: A California company known as Solazyme received $22 million in public taxpayer dollars through the Department of Defense's research arm, called DARPA, to create a synthetic-organism-derived biofuel for the military, but the project was terminated when the technicians couldn't bring the price down below $149 a gallon. The company subsequently turned its attention to wrinkle cream. In other words, the great hope for synthetic biology to bring "green" fuels into the marketplace instead generated a new face cream paid for by the American taxpayer (the company eventually went bankrupt). When queried about how to bring the cost of the fuel down, one of the biotechnologists involved in the project said the key was to make cheaper sugar.

Last comes the question of ethics. Gene editing and CRISPR-Cas9 made headlines in late 2018 with the surprise announcement that a Chinese scientist, He Jiankui, had edited the genomes of twin girls in vitro, altering their DNA. His goal was to see if he could create one baby with resistance to HIV infection by eliminating a protein called CCR5, which plays a key role in immune system response. If successful, the genetically modified child would have the opportunity to pass along her genes to subsequent generations. This self-replicating aspect was one of the many alarm bells raised by this development (the baby girls were born healthy). Apparently left out of He's experiment were regulators, overseers, policy makers, and fellow scientists, though he claimed to have gained permission from two colleagues. In the subsequent media and scientific firestorm, he was criticized for poorly designed protocols, a lack of transparency, and a failure to meet ethical standards for protecting the welfare of the subjects.

His rogue experiment drew a rebuke from Feng Zhang, a research geneticist who developed the CRISPR-Cas9 system. "Although I appreciate the global threat posed by HIV, at this stage the risks of editing embryos to knock out CCR5 seem to outweigh the potential benefits," said Zhang in a statement, "not to mention that knocking out of CCR5 will likely render a person much more susceptible for West Nile Virus. Just as important, there are already common and highly-effective methods to prevent transmission of HIV from a parent to an unborn child."[26] However, most scientists working in the gene-editing field rebuffed calls for a moratorium on future work on human embryos, insisting that while such a procedure might appear unethical—prompting fears of "designer" babies—the science has advanced enough to consider treating genetic disorders, such as cystic fibrosis and hemophilia, with CRISPR technology. In an interview after the announcement of He's experiment, George Church, author of *Regenesis* and a leading research bioengineer, voiced his support for the general approach, though not the way that He went about his work. "I'm not saying [there will] never be an off-target problem," Church told a reporter. "But let's be quantitative . . . we have pigs that have dozens of CRISPR mutations and a mouse strain that has forty CRISPR sites . . . and there are off-target effects in these animals, but we have no evidence of negative consequences."[27] Not yet.

The tantalizing vision of generating secure profits and new resources from microbes that will not form labor unions, get sick, or ask for maternity leave seems like a great win for venture capitalists, but is it a long-range win for the planet they live on? Is it right to overturn centuries of culture- and time-tested agricultural traditions in the rush to embrace a radical and unregulated technology? We must initiate and sustain robust ecological, cultural, and economic conversations regarding the context of these ventures. Instead of viewing culture, health, and economic equity frameworks as hurdles that technology companies must overcome and dominate, we need to assess the technology as it affects these systems. We may come to find that science for measurement and modeling is a far higher calling than science focused on manipulation aimed to redesign nature.

California cotton seeds extracted during ginning.

CHAPTER FIVE

Implementing the Vision with Plant-Based Fibers

So far we've discussed the environmental, social, and health-related costs of clothing produced by the current industrial system. We've highlighted the perils, both known and unknown, of proposed remedies to the system's global impacts involving GMO technology and synthetic biology; we've explored the idea of restoring regional fiber-and-dye systems known as fibersheds; we've gone on a personal journey into the Northern California fibershed and the subsequent formation of an educational nonprofit organization; and we've analyzed the multiple benefits of restoring the carbon cycle in soils, including its potential for drawing down hazardous buildup of carbon dioxide in the atmosphere. In the next three chapters, we will turn to the fundamental components that bring a fibershed to life: plant and animal fibers, milling infrastructure, and the relationships that hold everything together. We'll start with the most popular fiber on the planet and look at the challenges and opportunities it presents for communities that have the established climate and conditions to grow it—cotton. Cotton accounts for approximately 30 percent of the fibers we wear today, and is by far the most popular plant fiber utilized by clothing brands—the closest second is tree-based textile fibers at 6 percent of our fiber "diet."

Cotton

On the southern tip of my fibershed region are six major counties—San Joaquin, Fresno, Kern, Kings, Merced, and Tulare—all located within the San Joaquin Valley, an agricultural powerhouse famous the world over for its raisins, nuts, fruits, and vegetables. But far lesser known is

California cotton growing to maturity at Bowles Farm through the care and skill of Gustavo Urenda.

the large amount of conventional white cotton fiber grown in these counties. In fact, California is the fifth-largest producer of cotton in the United States, harvesting over 250 million pounds annually, and the San Joaquin Valley is nearly the *sole* source of the nation's Pima cotton, a varietal that produces longer fibers than its acala cotton cousin. Cotton grows successfully in warm, temperate climates with long growing seasons, similar to the lion's share of crops grown in drier parts of our state, requiring external irrigation. Thanks to the climate, water contracts, and level of long-term investment that farmers have put into cotton planting, harvesting, and ginning systems, California continues to be a significant contributor to the global cotton pool. But due to the organization of the global cotton commodity markets, a very small number of Californians are actually wearing cotton that is grown in our state. Only recently has some of our cotton been made consistently and transparently available to residents as a wearable textile through companies like Vreseis and California Cloth Foundry. These projects show promise for creating accessible apparel options from our region's cotton farms, and yet they represent a fraction of the cotton supply that is grown in our region.

Cotton production has slowly but steadily declined in California as the global demand for Pima cotton has dropped. The rise in production

California cotton being harvested, ginned, and baled.

of low-quality cotton in China, the shift to synthetic fibers by designers, and the squeeze on local water resources have all combined to threaten an industry that employs over 25,000 people directly in California and another 150,000 indirectly. One of many reasons that we do not yet have an opportunity to wear California cotton is the lack of functioning combing or spinning facilities within the state. The fiber is often sent overseas for milling—or at best is sent to the last few remaining cotton mills on the East Coast. Global-scale commodity systems have done a very good job of making it harder and more expensive to purchase a locally grown cotton T-shirt than it is to drive to a box store and purchase an equivalent garment grown and constructed on several different continents.

While cotton acreage has fallen to unexpected lows in the last nineteen years, the San Joaquin Valley (and California as a whole) still has the largest swath of irrigated land in the nation and accounts for one-fourth of total applied acre-feet of irrigated water in the United States. This same acreage of irrigated farmland has increased its production from thirty-three million tons in 1964 to seventy-three million tons in just over four decades, and water efficiencies have allowed communities to grow more with less.[1] However, the reality of the San Joaquin Valley is that it's a historically dry landscape; its farming systems consistently require water inputs from other watersheds and from currently deplet-ing aquifer systems.

In our summer dry season, approximately 2.5 acre-feet of water is used per year per acre for a yield of three to five bales of cotton (a bale is five hundred pounds). According to the California Cotton Ginners and Growers Association, it requires 256 gallons of water to produce a T-shirt. However, this water usage is noted by the association to be much less than the majority of other crops grown in cotton regions. And yet the benefit of California cotton compared with cotton crops grown in other warm climates is that farmers in California must pay for water use *and* they have access to technologies that increase the efficient use of water. In fact, through the Sustainable Groundwater Management Act, Califor-nia is moving groundwater into a regulatory framework where it will be measured and monitored for recharge. Pesticide use and herbicide use are publicly documented, and anyone can evaluate farmers' usage pat-terns, marking a unique level of regulation and transparency that provides end-users the information they need for deciding how and if they want to engage with cotton systems.

Establishing a holistic and climate benefiting set of practices for cot-ton in our fibershed is still a challenge, considering that we are in a highly prized set of growing climates with cotton fields that have been developed and managed since the early part of the twentieth century. Farming prac-tices have been honed and perfected, thus making any sudden changes to these systems is not simple. Conventional cotton seeds are coated in neonicotinoid pesticides and fungicides, making it very difficult to initiate and even trial regenerative and organic agriculture methods in our region. It has become quite difficult for new farming operations to obtain the necessary quantity of organic white cotton seeds to start a holistic farming model sized for economic efficiency in our region today; there is currently no organic seed bank for white cotton in our state. The closest organic cot-ton farmers to us (several states away) grow just enough seed for their own purposes, and so there is proving to be a growing need for regional organic seed banks and the infrastructure to gin organic cotton regionally.

> There is both room and opportunity to support a healthy transformation of California cotton systems.

There is both room and opportunity to support a healthy transforma-tion of California cotton systems. It will require that we not turn our backs on these farming communities even though they might be reliant on sys-tems at this moment that we know carry too high an ecological cost to be able to continue well into the future. We need to embrace these farms and support them to stay in business *and* engage in climate benefiting practices. In time, the hope is that by seeing for themselves the economic viability of regenerative agriculture systems, farmers will have all the incentives and market signals necessary to reduce and eventually eliminate their reliance on socially and environmentally expensive synthetic compounds.

One of the key levers for implementing an agricultural system that restores ecological systems to balance is building regional processing and milling systems whose business structure is designed solely to pay a premium for raw materials grown in a manner that supports our land-use goals. If a farmer or rancher is able to easily access local milling infrastructure and receive a price premium for sending raw goods to that facility, positive changes to on-farm management will follow. It's critical that we begin valuing multiple ecological and community-health-providing indicators within our business contracts and not just rewarding growers for high yields. In addition to the potential on-farm benefits, increasing the quality and quantity of local fiber processing will also diversify the livelihood potentials in rural communities, which will have rippling and overarching societal value. Pairing with a conscious and nuanced-thinking customer base that values local fiber products in the same way it values local, organic food is essential; and none of this works without the designers, weavers, sewers, educators, and other textile professionals, who are all vital to the journey from field to store. In the words of Peggy Sue Deaven-Smiltnieks, an award-winning clothing designer in Toronto, Canada, who has committed to the regionally based natural fiber movement: "No farms, no fashion."

A Brief History of Cotton

Cotton use began in small-scale prehistoric agricultural settings, where it was grown in an integrated manner with other crops. Our cotton-growing ancestors would never have imagined the lengths humans would go over the subsequent centuries to satiate a desire for the fine garments made from this plant material. The underpinning of our modern global economy and the structures that we now accept as normal, from international manufacturing epicenters to large-scale monocropped agriculture and the large-scale fast-paced consumption of textiles, can be traced back to technological revolutions in combination with human slavery, a system that was instituted as a means to increase access to cotton textiles. In my fibershed we work to renew the appreciation and understanding of the sacredness of cotton. The plant itself is beautiful and makes an incredible complement to food and natural dye gardens, homesteads, and small farms. Bringing cotton into our lives in a botanic capacity, not just by wearing it, will help us reconstruct a culture of appreciation for this venerable plant.

A member of the mallow family, the progenitor of modern cotton plants emerged on Earth between ten and twenty million years ago, evolving into small, frost-intolerant, perennial trees that grew hard seeds covered with short soft hairs. Domestic cotton (the word is derived from the Arabic *qutun*) grows on shrublike plants and is the soft, lintlike fruit, or *boll*, that shrouds the seeds after flowering. Classical breeding over the centuries turned cotton into an annual plant and increased the size of cotton seeds—which in turn increased the amount of lint fiber available for spinning. The length of the fiber is called *staple*; it can be short or long depending on the plant species and other factors. Long staples are preferred for high-quality textiles, such as Pima cotton, while short staples are widely used in everyday garments and household products. In the San Joaquin Valley, ideal growing conditions have also allowed the development of a variety of short-staple cotton that has the feel of silkier Pima cotton. In all species and varieties, seeds must be separated from the fiber before the cotton can be used, a process that is both tedious and time-consuming.

The earliest archaeological evidence of cotton's domestication has been found in the Indus Valley of Pakistan and dates to approximately seven thousand years ago. It is closely affiliated with the rise of city-states in the region and the development of specialized craft trades, a critical step on the road toward increased social complexity. The species found in Indus Valley is *Gossypium arboreum* (tree cotton), and it steadily spread around the world through trading networks, eventually becoming an integral part of the economies of early civilizations, particularly in the Indian subcontinent where growing conditions were ideal. Evidence of cotton textiles has been found in ancient sites in Iraq, Greece, Iran, North Africa, and China. A distinct but related species, *G. herbaceum* (shrub cotton or levant cotton), originated in Africa; although it was domesticated it was largely neglected in favor of its cousin.

In the New World two species of cotton developed separately from their Old World relatives and now account for more than 95 percent of all production globally. *G. hirsutum* (known today as upland cotton) was first domesticated more than four thousand years ago in coastal Mexico and on the Yucatan Peninsula. Cultivation spread over time, and cotton cloth—colored with indigo, cochineal, and other natural dyes—became an important luxury item for nobles and kings as well as a highly coveted trade item among prehistoric Mesoamerican societies. Domestication of

Hand-harvesting naturally pigmented Pakucho cotton that is at least five millennia old in the upper Amazon regions of eastern Peru. *Photo by Dr. James M. Vreeland, Jr.*

the fourth cotton species, *G. barbadense* (called Pima, Egyptian, or Sea Island cotton), happened among prehistoric cultures on the coast of Peru, where it was used to make fishing nets and other objects associated with marine agriculture. Eventually it spread inland and was incorporated into numerous textiles, including clothes and bags. Similar to its Mesoamerican cousin, Peruvian cotton went through a lengthy breeding process by which humans selected plants with the traits they preferred. Today the annual crop has many more seeds than its ancient relatives, all covered with plentiful long, soft fibers.

The early cotton farmers who lived scattered about the planet had several farming traditions in common. They raised cotton alongside other useful crops and grew just enough to serve themselves and their families first. As cotton became increasingly popular over the ensuing millennia and spread around the world, the crop continued to be grown, harvested, and processed by small landholders—a tradition that can still be found today in small pockets of people. The only large centers of population that did not cultivate cotton as a crop were in Europe, where textile makers had refined their skills around the fibers that they had access to within their own bioregions. (Europeans relied heavily on wool and linen, which persisted as the raw materials of choice for textile production for millennia in the colder and wetter climates.) The first sign of cotton making its way into modern-day Europe was through Africa into southern Spain a thousand years ago, into the Islamic cities of Cordoba, Granada, and Barcelona, where it was manufactured into textiles.

The cotton industry developed slowly for the next thousand years until a series of events in the late eighteenth century created the conditions for a radical shift in social, technological, and economic systems driven primarily by the increased production and consumption of cotton textiles. Engineers in the United Kingdom pioneered technological advancements such as the flying shuttle, invented by John Robert Kay, which considerably sped up Britain's looms. Spinning efficiencies were developed in close succession, beginning with James Hargreaves's invention of the spinning jenny. A further spinning advancement, the water frame, was developed by Richard Arkwright, followed by an even more efficient machine known as the spinning mule, invented by Samuel Crompton. The efficiencies of these technologies were staggering. Mills in Britain could now spin a hundred pounds of raw cotton in 135 hours, whereas in India the work to spin the same type and quantity of yarn required 50,000 hours.

Advances in technology continued to quicken the pace and scale of cotton textile production. In 1833 Walter Hunt invented the lockstitch

Eli Whitney and the Cotton Gin

Ginning is the process of separating cotton seeds from the surrounding fiber, and for millennia it had to be done by hand. In 1794 inventor Eli Whitney created a mechanical cotton gin while living on a slave plantation in Georgia. Within ten years this de-seeding device increased farm production tenfold, reinvigorating the slave trade at a crucial moment in American history (eventually provoking the Civil War). Where a single worker could remove seeds at a rate of one pound of cotton per day by hand, Whitney's invention could remove seeds from fifty pounds in the same amount of time. The South's economy grew dramatically as a result, causing the now very profitable crop to be called King Cotton. By 1860 it is estimated that one in three residents of the South was a slave. The North's economy transformed due to the invention, since most of the textile mills were located in New England, and so there were rapid advances in transportation technology, including the famous Mississippi steamboat, which abetted the expansion of the cotton industry in the United States.

This high-throughput system propelled cotton milling systems in Europe as well. By 1800 Britain was spinning fifty-six million pounds of cotton a year, ten times the amount that had been spun just twenty years prior. There were widespread collateral effects of this scaling-up, including the vast expansion of colonial collateral damage in areas of the world where large amounts of raw cotton could be grown. England was one example, first through private enterprises such as the East India Company and then through direct action by the government, which occupied the Indian subcontinent—the source of the finest cotton fabric in the world.

Another Eli Whitney invention had a significant impact on society as well. After receiving a contract from the US government to produce ten thousand rifles in two years—in an era when guns were essentially handmade by craftsmen—he devised a manufacturing process that utilized standardized and interchangeable parts, vastly increasing the speed of production. The guns could also now be made by relatively low-skilled workers, setting the stage for the industrial model of mass production that came to dominate the American economy.

Whitney's technologies bolstered both the slave trade and the mechanized systems of violence required to keep colonial interests in power. But technologies such as these are applauded in mainstream texts because they are credited with bolstering economic advantages for those who wrote much of published history. There remains little discussion to this day regarding the relationship of the cotton gin and interchangeable parts for guns to the historic and ongoing social, economic, and political costs our entire society has experienced due to the designing of, investment in, and dissemination of these machines and weapons. As we move ahead to develop cleaner and equitable fiber and dye systems, it is of great value that we first take the time to understand the historic roots of oppression within agriculture and within the way technology was and continues to be used to centralize power, so as not to repeat our mistakes.

sewing machine, which increased the incorporation of cotton fabric into the garment industry. In 1885 the cotton planter was patented by Henry Howell. Later, over the course of many years, the mechanized cotton picker was invented by Daniel Rust, who received his first patent in 1933.

Meanwhile new lands in North America continued to be forcibly opened up to settlers throughout the late nineteenth and early twentieth centuries, resulting in indigenous communities being pushed deeper into the marginal recesses of the nation. The development of these systems of production laid the foundation for a new form of global capitalism that disaggregated centralized systems of agriculture, manufacturing, and end-users and spread them to far-flung continents, and we have been largely reliant on this disaggregation to the present day. Large tracts of land have been systematically taken from indigenous tribes through European colonial conquests in these regions and then forcefully populated with slaves to work on the plantations. Even today within remote regions of Ethiopia's Omo Valley, pastoralist societies living with cattle and goat herds are being displaced by cotton plantations, and rivers are being dammed for irrigation water.

In his seminal history of global cotton titled *Empire of Cotton*, Sven Beckert stated that the combination of imperial expansion, slave labor, machine technology, and increased productivity launched the Industrial Revolution with all its deleterious ramifications, including labor abuses, a host of social and health problems, and staggering inequality. Beckert called what happened in the cotton fields of distant lands "war capitalism" because it involved the armed theft of land, ruinous exploitation of natural resources, and the forced subjugation of indigenous peoples. "Europeans united the power of capital and the power of the state to forge, often violently, a global production complex," he wrote, "and then used the capital, skills, networks, and institutions of cotton to embark upon the upswing in technology and wealth that defines the modern world."[2] Particularly troubling was the rapid and vast wealth accumulation that took place among a narrow band of elites in Europe and America. "It was the beginning of the vast divides that still structure today's world," Beckert said, "the divide between those countries that industrialized and those that did not, between colonizers and colonized, between the Global North and the Global South."

Cotton Today

The multiple issues with cotton production continue to be compounded in the modern era by environmental, labor, and commodity pricing troubles, all of which will intensify under the stress of climate change. As described earlier in this chapter, much of the water used in cotton cultivation arrives at the plant's roots via irrigation, which stresses scarce water resources and adds to depletion of groundwater tables, rivers, and lakes. Five of the top nine pesticides used on cotton are known carcinogens, and all nine are classified by the EPA as dangerous.[3] In the San Joaquin Valley, it is estimated that as much as 25 percent of a pesticide sprayed from a crop-dusting airplane never hits the crop target. The remainder can drift for miles, falling on neighboring crops (including organic ones) and farmworkers. In 2014 approximately 94 percent of all cotton acreage in the United States was planted in GMO varietals, of which there are at least fifty-four different types.[4] Cotton seeds are also fed to livestock in the dairy industry. However, in 2018 the USDA approved a genetically modified seed that is digestible by humans, engineered using DNA insertion techniques (in other words, synthetic biology). One of the lead researchers, Keerti Rathore of Texas A&M University, told an interviewer that his goal was to create a food product that was safe to consume (conveniently ignoring the herbicides and pesticides used in cotton farming). "A lot of these countries that do suffer from malnutrition are also cotton producers," he said. "So I think those countries may benefit much more from this technology."[5]

> In the San Joaquin Valley, it is estimated that as much as 25 percent of a pesticide sprayed from a crop-dusting airplane never hits the crop target.

Another major challenge for cotton production today is its long history of forced labor at harvesttime. Historically this labor was done by slaves, and illegal slave labor can still be found in many parts of the world, according the US Department of Labor, including nine countries that produce 65 percent of the world's cotton: Benin, Burkina Faso, China, India, Kazakhstan, Pakistan, Tajikistan, Uzbekistan, and Turkmenistan.[6] Uzbekistan is the most notorious violator—every year the government forcibly mobilizes a million of its citizens who hold regular jobs, such as teachers and doctors, to go into the fields during harvest to pick cotton.[7] Until 2012 this labor pool included many children, but pressure from advocacy groups and governments in countries where Uzbek cotton is exported successfully changed the policy. One consequence of this campaign was the creation of the Responsible Sourcing Network, a project of the nonprofit organization As You Sow dedicated to ending human rights abuses and forced labor associated with products we use every day.[8]

Given the widespread challenges confronting cotton production and manufacturing today, it is a fair question to ask what a soil regenerating or Climate Beneficial system of cotton might look like.

Cotton and the Aral Sea Catastrophe

In 1988 the Soviet Republic of Uzbekistan was the largest exporter of cotton in the world. It came at a terrible environmental cost, however—the destruction of the Aral Sea. Shortly after World War Two, the Soviet government began diverting water from the two major rivers that fed the Aral Sea in order to transform the surrounding desert into a cotton-growing powerhouse. The project expanded in the 1960s using poorly constructed irrigation canals that wasted as much as 70 percent of their water before it reached the fields. Poor water management on the farms resulted in a doubling of withdrawals from the two rivers by 2000. Meanwhile cotton harvests rose steadily. As the export market developed and demand for Soviet "white gold" grew internationally, government officials insisted on greater yields from the farms and their laborers.

Once the fourth-largest body of water in the world, the Aral Sea steadily shrank in size. By 1998 its surface area had diminished by 60 percent and its water volume by 80 percent. At the same time the Aral's salinity level rose dramatically, surpassing levels in Jordan's famous Dead Sea by 1990. The salt killed off native fish populations and destroyed the fishing industry, which once employed forty thousand people. Continued drying soon split the Aral into two separate bodies of water. In 1991 Uzbekistan gained independence from the Soviet Union, and the country's dictator, Islam Karimov, demanded increased cotton production in the Aral Basin to boost the nation's struggling economy. To offset faltering yields, larger and larger amounts of chemical herbicides, pesticides, and fertilizers were used, further damaging natural systems. The rate of shrinkage sped up, aided by frequent droughts, culminating in a declaration by NASA scientists in 2014 that the Aral Sea was functionally dead.

It wasn't a surprise to former Soviet-era engineers, one of whom said the sea was "nature's error" and would have dried up anyway. The exposed salt bed has created hazards of its own. The land surrounding the Aral is heavily polluted with industrial chemicals, and windborne toxic dust storms plague the region, causing elevated levels of cancer and tuberculosis. Child mortality rates have soared, and the list of troubles goes on and on.[9] What happened to the Aral Sea is a shocking tale of ecosystem destruction at the hands of industrial agriculture and a disturbing example of human hubris and greed.

Climate-Friendly Cotton Projects

I've already introduced several times the work of Sally Fox, who has a long history of both wearing and growing her own cotton and has spent her later life providing her product to artisans, including hand-spinners, weavers, and brands. Sally lives in an agricultural valley near Sacramento and runs a farm that is certified organic and biodynamic. Her geographic distance from other cotton growers provides her the freedom to carefully breed varietals, all of which are color-grown species. For as long as colored cottons have existed (many millennia), they were and, in many cases, remain a short-staple crop, one that historically has required the skilled hands of well-practiced spinners. Color-grown varieties were commonly grown prior to the industrial era, but the machines that revolutionized cotton were designed to manipulate the fibers of longer-staple white cottons. Industry preferred white cotton so they could uniformly dye the plant fibers with their newly invented synthetic dyes. As time passed white cotton became the dominant fiber of choice for the industry, and so we have a forgotten that cotton naturally grows in many beautiful different hues. In fact, the colors are so rich in tone that they require no dye at all—and yet, if you do care to use dye, they require much less, which makes them an incredible option and solution for our water resource usage and synthetic dye effluent concerns.

Along with growing colored cotton, Sally has implemented a variety of soil building farming practices. She co-plants black-eyed peas with her cotton to attract the pests and lure them off her cotton crop. She rotates cotton with an heirloom variety of wheat and grazes her wheat and cotton stubble with a flock of Merino sheep, all of which helps build soil biology, sequester atmospheric carbon, and enhance the water-holding capacity of her soils. Sally likes to remind people that cotton is similar to any other plant—it needs healthy soils, an adequate supply of water, and sunshine to thrive. There is nothing intrinsically "problematic" about cotton except the way it has been traditionally farmed. She doesn't use any pesticides, insecticides, or synthetic fertilizer to raise her crop, and she needs to water it only once every two weeks—utilizing less water than many of her vegetable- and fruit-growing neighbors.

Approximately one hundred miles north of Sally's farm, an important cotton farming and research project is in the early stages. A new program at Chico State University, directed by Dr. Cindy Daley and known as the Center for Regenerative Agriculture and Resilient Systems (CRARS), aims to research and demonstrate regenerative agricultural practices that both restore

Yarns that Sally Fox designs and sells straight from her farm.

Sally Fox: breeder of color-grown cotton, scientist, farmer, and textile designer extraordinaire, owner of Viriditas Farm in Capay Valley, California.

and enhance the resiliency of living systems. One of the crops they've included in their research is cotton. Working with Dr. Tim Lasalle, a soil health advocate, and Dr. David Johnson, a molecular biologist at New Mexico State University, the CRARS in collaboration with regional cotton growers and Fibershed (the nonprofit) is exploring the use of fungal-dominated compost extracts and multispecies cover cropping to improve soil biology, carbon sequestration, and crop yield.

Johnson believes too much research over the years has focused on poor soils or soils that have been damaged by chemical inputs, while not enough has been done on healthy plants in healthy soils. His mission is to change that trend. His work on cotton, corn, and other crops indicates that enhancing the microbiological community improves crop productivity, restores soil water-holding capacity, and improves drought resistance while also providing a long-term sustainable approach to agriculture. In addition, Johnson has analyzed how carbon restored to soils will eventually outperform energy- and capital-expensive carbon-capture technologies for atmospheric carbon reduction. He has coined the term BEAM for his methodology, which stands for "Biologically Enhanced Agricultural Method." The goal is to achieve a bacterial-fungal balance in agricultural soils, which generally means moving away from the bacteria-dominated systems that are typical of chemically heavy industrial farming systems. This requires a holistic approach to a farm or ranch operation, including the integration of cover crops, compost, and livestock together where practical. The key to BEAM is bringing back fungal populations and restoring their place in the soil-food web. It can take as little as one season of cover crops to get fungi back into the system, though two or three seasons are usually needed for the full effect.

One critical component of the holistic system that restores bacterial-fungal balance, Johnson believes, is the application of fungal-rich compost extract that is cured over time in a vertical composting system Johnson and his wife invented, which they call a *bioreactor*. It is simple and inexpensive to construct (for as little as forty dollars). Once it has been filled with organic material, such as wet leaves, you don't need to turn the compost. With proper air circulation, daily one-minute waterings that can be automated, and one year of cure time, the compost extract becomes a substance that is teeming with life. Once it is complete, the fungi-laden compost extract can be applied to a field, injected into the soil, or used to coat seeds before they are planted. One of the experiments that Johnson did in New Mexico was to apply the compost to a cotton crop that was planted into soils that had grown cover crops for multiple previous seasons. The plants grew up to six feet high, and when they were harvested, Johnson reported they netted five bales per acre, more than double the average cotton yield in the area, which typically receives less than five inches of rainfall annually. The goal of the CRARS project is to replicate Johnson's experiments and rigorously monitor and quantify their potential.

To rebuild our soils and implement Climate Beneficial agriculture for the long run, especially in semi-arid environments, cotton farmers will need to adopt models that consider how to maximize the use of the precipitation that falls within their own watersheds. Climate change has put a time stamp on our irrigation systems, one that will greatly limit and could eventually eliminate irrigation that relies upon snowmelt from distant watersheds. There are cotton-growing systems not too far afield from our region in the West that work with the precipitation rates in their own watershed as a means of covering as much of their water budget as possible, but these are areas that are known to receive summer rains that can nourish the cotton during its growth cycle, unlike the dry summer conditions in California. In Texas, for example, thousands of acres of organic cotton are grown annually by a cooperative of farmers who rely heavily on dry farming techniques. Founded in 1993 and headquartered in Lubbock, the Texas Organic Cotton Marketing Cooperative is the source of the majority of organic cotton grown in the United States (in 2016 it grew 84 percent of it). Its forty members, each family farm certified organic by the USDA, plant up to twenty thousand acres of cotton each year, netting between ten and seventeen thousand bales, nearly all of it grown without the benefit of irrigation. Each bale of organic cotton is tracked from the farm to the customer, which provides critical supply-chain transparency. Marketing services are provided by the cooperative, whose members grow other organic crops, including wheat, corn, soybeans, milo, peanuts, sorghum, black-eyed peas, and watermelons. The ginned cotton seed is also marketed to organic dairies for feed.

Another inspiring example of a rain-fed cotton-growing system can be found in India, the historic center of global cotton production and still a major player, producing 20 percent of the world's cotton (second only to China). In 2007 a small group of friends gathered in Chennai, a city on the Bay of Bengal in southern India, to discuss their interest in reviving organic farming and promoting indigenous foods as a way of preventing further suicides among farmers, a crisis that has ravaged India for years. What emerged from the discussion was a nonprofit store that connected consumers to local organic farmers who use traditional and sustainable agricultural methods. That effort led to the innovative Organic Farmers Market, by which families in eighteen boroughs around Chennai make a room available in their homes twice a week so that neighbors can pick up freshly delivered organic produce. The food isn't certified organic, however, which would involve India's famously complex and stifling bureaucracy. Instead, buyers and sellers have a direct relationship through the market, relying on one-to-one trust that their food was grown ecologically.

In 2013 the same group of friends gathered again, hoping to re-create their success, but this time they focused on making a dent in the high rates of farmer suicide. They knew the majority of suicides occurred in India's cotton belt and were linked to the high cost of herbicides and pesticides required to grow the genetically modified Bt cotton that farmers had been convinced to grow. The rising costs of the GMO seed and the chemicals, combined with diminishing yields, stagnant cotton prices, and increasing resistance to the chemicals by pests, were driving many poverty-stricken farmers to kill themselves, sometimes with the very poisons they used on their crops. The group decided to explore the organic cotton value chain to see how they could improve rural livelihoods and reduce ecological harm. They named their effort

Sally Fox's cotton was used in several designs for the one-year wardrobe. Rebecca Burgess stands within her small indigo farming project to elucidate the material connections between soil, plant, and skin. The hand-knit tops were designed and made by Heidi Iverson, sourced from Fox's Palo Verde cotton yarns that were hand-dyed by Burgess from the indigo in the pictured field. The skirt was sewn by Kerry Keefe with Fox's buffalo brown cotton jersey that was provided by Fox from her archive of personally designed textiles.

Tula, which is the Sanskrit word for "cotton" (it also means "balance" or "equilibrium"). Their goal was to restore a balance in the cotton value chain, which had been originally destroyed by British imperialism and magnified by the chemical-dependent Green Revolution that began in the 1960s.

As the Tula experiment unfolded, the team was urged to support a revival of an indigenous variety of the species *Gossypium arboreum*, a short-staple cotton adapted to the local climate, including its insects, that does not require irrigation. This proved to be a critical decision. The Tula team worked with local communities to identify the particular strains of cotton unique to each locality, sometimes coming across elderly farmers who had saved old seeds and sometimes finding seeds in local agricultural universities. Tula encouraged farmers to use traditional Indian methods to grow the seed, including planting a polyculture in a field, employing chili, corn, onion, tomato, and other crops side by side with the cotton. To make this effort financially viable to the farmers, Tula pays a higher price than what a comparable amount of Bt cotton would fetch on the market. Today Tula buys more than two hundred quintals (one quintal equals one hundred kilograms) of cotton annually from nearly one hundred farmers, who, on average, own less than an acre each. The Tula organizers and investors next turned their attention to the spinning, weaving, and dyeing parts of the cotton value chain in an effort to revive an ancient tradition called khadi while at the same time drastically reducing the vast amounts of energy and water that used to produce fabric under the current industrial model.

The Tula experiment has been successful. Revival of the region's traditional cotton industry is under way, and khadi institutions are growing once more in various places in India. The original financial investments are being paid back; the number of cotton-based livelihoods linked to Tula has grown to more than three hundred. The effort won't stop there, however. Work needs to be done to help indigenous seed banks and seed breeders meet rising demands from farmers. There is also a critical shortage of skilled spinners and weavers to be filled. Additional markets in the region for the organic cotton need to be developed (the product is not exported), and efforts like Tula need to be replicated. But regardless, it is a very promising start.

> **Work needs to be done to help indigenous seed banks and seed breeders meet rising demands from farmers.**

Flax growing at the Chico Flax project.

A Flax Experiment

In the foothills of our Northern California fibershed in the southern Cascade mountain range near Chico, California, a new group of growers has come together to grow flax, which is the source fiber for linen. Organized by Sandy Fisher and her husband, Durl Van Alstyne, Chico Flax LLC began by successfully raising an experimental crop of flax during the winter, the time of year that California normally gets most of its rain. The project aims to develop a sizable seed bank for the region and has brought together students and professors from Chico State University, gardeners, interested textile artisans and designers, and others to develop the skills needed to scale up this project within their community. Their ultimate aim is to invest in a milling system to mechanically process their fibers and develop their flax into a regionally branded cloth. As with other natural fibers, turning flax into linen requires a community effort involving farmers, spinners, dyers, and weavers, not to mention designers, educators, and machinists. Sandy has been a professional weaver for more than thirty years and as a result was able to reach out to a well-developed weaving community, including members of the Mount Lassen Fiber Guild, which was founded in 1955.

The one-year wardrobe challenge included inspiring pieces crafted by design school graduates from California College of the Arts. Zara Franks was connected to the challenge through her professor Lynda Grose. Franks knit this shirt from Sally Fox's color-grown cotton yarns.

Sandy and her husband were motivated to start Chico Flax LLC when they heard about a terrible fire at a textile factory outside Dhaka, Bangladesh, in late 2012 that killed 117 workers. Shortly after the tragedy happened, Sandy received a phone call from a local peace activist who wanted to know how to source her clothes in the Chico area. A dialogue ensued as Sandy and the activist tried to answer the question: "What can local weavers and wearers do as an answer to this horrible situation?" They found inspiration in Mahatma Gandhi's belief that his fellow Indians could take a big step toward self-reliance (and throw off the yoke of British rule) by making their own textiles locally. After some further discussion, Sandy and her husband decided to take the plunge and engage the local community to help produce a fabric that would represent their home while having the potential to expand beyond the upper Sacramento Valley. They chose linen. Flaxseeds were purchased and a small plot of land secured for the experiment. Instructions on how to grow the plant properly were found in a publication called *The Big Book of Flax*. They crossed their fingers. A few months later they harvested their first crop—and Chico Flax LLC was under way.

The Pacific Northwest Fibershed leadership is pioneering flax production in their region (I will provide more detail on their project in chapter 7). The relationship to flax in the Pacific Northwest region dates back to the mid-19th century. The project is currently focused on collecting results to find the best agronomic practices for optimal fiber quality and yield and working to certify the community's future processing facility to the Global Organic Textile Standard for manufacturing, which will support incentives for farmers to grow organic flax.

Flax is a tall, thin plant that grows to maturity in less than hundred days and produces a bluish flower with a delicate shape. Flax, or *Linum usitatissimum* (most useful), is a bast plant, which refers to the fibers that are produced in each stalk, helping the plant stand upright and not fall over as it grows a head of heavy seeds. Flax's long, straight, smooth fibers are durable, are anti-microbial, and have a beautiful weight and drape to them when made into linen. At harvesttime the plants are pulled by their roots from the ground if possible, not cut, which guarantees the longest unbroken fibers. Next comes the retting process, by which the fibers in the harvested flax stalks are exposed to water, such as submersion in a pond, which softens them through microbial action, a necessary step before their eventual use as textile fiber. The highest-quality fibers come from stalks left in a field for as long as a month, a process called dew retting, which can produce a darker color of fiber. There are other retting methods, including chemical ones, which can speed up the softening process. Once dry, flax then goes through a three-step procedure that separates the fibers from the rest of the stalk (called *breaking*, *scutching*, and *hackling*) in order to prep it for spinning into yarn that can be made into a variety of linen products. As with cotton, short plant fibers can be used for coarser material while the longer fibers make silkier products. Linen is absorbent, quick drying, and cool feeling. It takes up color more easily and shrinks less than cotton.

Flax could be an ideal crop for polyculture farming. It grows well mixed with other annual crops, including cereals, grasses, and oilseed plants such as sunflowers, rapeseed, and soybeans. It is grown in the winter rainy months in California and Oregon and thus does not require much in the form of external irrigation, and its high germination rate means it has the potential to outcompete weeds, reducing or eliminating the need for herbicides. Flax's rapid rate of maturation means it is a perfect crop to follow a period of livestock grazing or a season of cover cropping. On a holistically managed farm or ranch, flax could be a part of a plan to sequester and store atmospheric carbon dioxide as described in chapter 3. Additionally, flaxseeds can be used as nutrient rich feed for cattle. Not only are they good for fiber, but flaxseeds have proven anti-inflammatory properties, useful for combating respiratory ailments in livestock.

> Not only are they good for fiber, but flaxseeds have proven anti-inflammatory properties, useful for combating respiratory ailments in livestock.

A Brief History of Flax

According to archaeological evidence, flax cultivation dates back at least thirty-six thousand years, suggesting that Paleolithic peoples may have been aware of the plant's utility for producing a wearable fabric. During the Agricultural Revolution flax was one of the first crops to be domesticated. Perhaps due to the laborious, time-consuming work needed to create it, fine linen became highly prized by the ruling elites in early civilizations, and ancient Egyptian mummies have been discovered wrapped in the finest linen. Flax's ability to grow in colder climates, as well as linen's capacity to dry quickly, meant that its popularity grew in the cooler parts of Europe over the centuries (alongside wool, discussed in the following chapter). In addition to clothing and other apparel, linen's uses expanded to include bedding, towels, curtains, tablecloths, religious vestments, sails, bowstrings, baking cloth,

A demonstration of flax hackling takes place
during a volunteer day at the Chico Flax LLC.

kerchiefs, money, luggage, sewing thread, and even shields. In the early eighteenth century, Ireland became the leading producer of linen in Europe, a distinction it held to the end of the Victorian era. In North America commercial production of flax fiber started early and grew steadily during the colonial period but declined quickly after the invention of the cotton gin by Eli Whitney and the ensuing expansion of cotton production. For a while flax fiber farms could still be found in North and South Dakota, Minnesota, and Wisconsin, but by World War Two production had virtually ceased across the nation. Growing flax for seed oil and tobacco paper, however, remained economically viable, and today flaxseeds are also a popular nutritional supplement for human consumption.

Recently there has been a resurgence in flax production in the US as well as the rekindling of knowledge about flax processing, including the development of the New England Flax and Linen Project. This agricultural research projects has instigated a body of work reflective of its geography and place and carries a sensitivity to what is most needed to help bring this time-honored textile fiber into the lives of modern-day wearers. In New England the group comprises a consortium of plant and environmental scientists, craftspeople, and historians. They held a symposium in 2016 that brought together skilled processors who had an expertise in farming, retting, spinning, and weaving, as well as field researchers and historians. New England is also home to the Marshfield School of Weaving in Vermont, a folk school that offers hands-on education in flax growing, processing, and weaving (for more information see the Resource Guide).

The knowledge base for flax in North America came with European farmers and other immigrants who had land-based livelihoods. These tradition holders often were self-sufficient in the manner in which they implemented their knowledge, which included growing the crop in small-acreage settings to have enough fiber to make their own bed linens, tablecloths, shirts, pants, and dresses. Skilled flax growers and processors had evolved over thousands of years in Europe, which may be in part why flax never received the same colonial attention from explorers and capitalists as cotton did. It is much more difficult to build exploitative ownership models with a material that everyone already has access to and already knows how to work. Perhaps if the Peruvians, Aztecs, and subcontinent Indians had been flax growers instead and Europeans had been the originators of the cotton legacy, then flax might have become the fiber of intrigue that defined imperial conquest.

Dew-retted and hand-processed flax is spun into yarn by a member of the Chico Flax project.

During an evening at the Marshfield School of Weaving not long ago, a linen-and-wool textile emerged from one of the student's looms. The cloth was then passed to a group of twenty or so of us who sat at a long table in a beautifully built barn. The textile was wetted in soapy water, and each of us held one section of the long tubular cloth, pounding it on the table twice and then rotating the cloth to the right. Pound twice, rotate, pound twice, rotate—the rhythm became a song whose melody emerged from several women at the table, the lyrics beautifully sung in Gaelic. As we sang, we worked, and the more we worked the more uniform the textile became. The process was topped off with a line dance, which provided us the opportunity to rise from the table and move about the barn. Traditionally, this process of wetting and pounding a textile is known as *fulling*. Machines now do this work for us, and this of course saves us time, but we do lose the practical culture that makes work a joy and provides us an opportunity to be part of a community. In another example of how fiber projects can build community, Chico Flax has drawn the support of Curtis DeBerg, a professor of business at Chico State University. He has given the project sound business advice and also directed some of his students to the budding flax endeavor, where they helped launch a crowdfunding campaign among other things. This link to the local university has expanded recently into new relationships with the agriculture and engineering departments, each eager to help. The ripple effects keep spreading outward, attracting bicyclists, bakers, retirees, veterans, and many others.

It takes approximately seven years—seven crops and their resulting seeds—for flax plants to become fully adjusted to their local environment. And since the project's seeds originated outside of California, this means the 2019 harvest will mark an important threshold for the four-acre Chico Flax farm as well as the overall project. A recent grant from the Department of Food and Agriculture's Healthy Soils Initiative, which supports carbon farming efforts around California, will be applied to developing a cover crop for the farm fields and the planting of hedgerows to attract native pollinators. There are plans afoot to convert an old barn on the property into storage for flax processing equipment. Meanwhile, Sandy, Durl, Curt, and other supporters are planning a party in the spring to celebrate the farm's transition to an indigenous crop. Step by step, the dream of creating community-wide, locally grown and harvested natural flax fiber is coming to a reality.

Hemp: Legal Once Again

Hemp was a very popular agricultural product in the early twentieth century, often grown in large amounts for use in the manufacturing of rope, paper, and other goods. Hemp textiles were widely used for clothing, home furnishings, shoes, and much more. But the adoption of drug laws in the early 1900s meant hemp was essentially banned from use (a condition formalized by Congress in 1970 when it passed the Controlled Substances Act). Hemp was bundled into the law due to confusion (fueled by economic interests) between the very different *Cannabis sativa* strains. Hemp has low concentrations of tetrahydrocannabinol (THC)—the chemical that makes a person high—whereas marijuana, a similar cultivar, has high concentrations of THC (20 percent and higher). Regulation of hemp moved out of the agricultural sector and fell to the federal Drug Enforcement Agency. But this was woefully misguided; there are over two thousand varieties of cannabis and over 90 percent have no druglike properties. Plants such as hops that are also in the Cannabaceae family contain levels of THC similar to that of hemp. Hops, however, are legal and used every day as an ingredient in beer.

In its effort to eradicate marijuana farms, the DEA destroyed wild and native hemp plants that grew in meadows, ditches, and along roadsides. Low-THC hemp is almost never grown in association with marijuana crops, partly due to its low value as a crop but mostly because of its ability to easily cross-pollinate with the high-THC cannabis varieties, which drastically reduces their psychoactive impacts (not something marijuana growers want). And yet this horticultural knowledge was somehow left out of our drug laws, and as a result we have lost decades of research and development time on a fiber crop that is known by many other cultures to create some of the strongest and most durable heirloom textiles in human history. I should say up front that I do not believe hemp to be the panacea it is often promoted to be, though it does offer an incredible opportunity to support our natural fiber systems.

Hemp has been grown and utilized by European and Asian cultures for centuries and has been bred into a multitude of varietals that have adapted to different elevations and latitudes. A hemp plant is sensitive to the amount of available sunlight, and many hemp breeders worldwide have been selecting for hemp fiber and biomass-dense plants that produce good stalk height prior to the summer solstice. Once the days begin to shorten and the plant flowers, the stalk will not continue to grow; whatever biomass the plant generates between planting and flowering will be the material farmers have to work with at harvesttime. Fiber-focused breeding efforts that favor fast-growing single-stalk plants have been under way on many continents for millennia, except of course within the United States, where farmers had to put aside the seeds, tools, and knowledge when the federal government initiated its ban.

In late 2018 approval of the Farm Bill lifted the ban and legalized hemp (or at least those plants that contain less than 0.3 percent THC). While this was very good news, there are still limitations imposed by the legislation. The previous Farm Bill allowed pilot research and marketing studies only on hemp (sometimes called industrial hemp) on small-acreage plots. While the new Farm Bill allows much broader-scale cultivation and eliminates restrictions on the sale, transport, and possession of hemp products, it still requires states to set up an oversight process by which hemp growers must be licensed and comply with all federal laws and regulations. States must also spell out what its punishments are for violators. In other words, while the new Farm Bill legalizes hemp, it doesn't allow for it to be grown as easily as broccoli or chard for commercial purposes.

> While the new Farm Bill legalizes hemp, it doesn't allow for it to be grown as easily as broccoli or chard for commercial purposes.

Why Hemp?

Hemp is a bast fiber plant, like flax, and while it has much to offer fibersheds, it requires a great deal of technological assistance to be truly useful in a textile. My first exploration into hemp textile processing was an exposure to the ancient technologies that people have refined over millennia as a means to turn plants into clothing. It was during trips to northern Thailand that I learned how the hill tribes produced narrow-width woven textiles, cloth that was produced for both subsistence and earning extra money at local markets. The hemp was grown in small plots and harvested by hand; stalks were gathered to dry in upright piles known as *stooks*, and then laid on the soil for several weeks after harvest to allow the stalks to dew ret. The microbiology in the soil breaks down the *lignin*, which is the glue-like substance that hardens and bonds the woody part of the plant, called the *hurd*, to the fiber. Once the lignin is broken down, the fibers can be more easily stripped from the stalk. When the stalks are perfectly retted, they are gathered from their prostrate position and stored upright so that the lignin breakdown process will slow.

Hemp fibers are long and strong, and when processed by hand, the full length of the fiber can be handspun on drop spindles and then woven. The resulting textiles from this region were washed and pounded so that they could further congeal and come together. The cloth was weighty for its volume and incredibly strong. These hand-processed textiles were good examples of how durable hemp textiles can be made through processes that rely on sun, water, microbes, and skilled hands. The low-tech approach in northern Thailand was similar to textile processing techniques that continue to this day in pocket economies in parts of Southeast Asia and eastern Europe. In the Loire Valley in France, a visit to the small town known as Mont Jean in August will bring a traveler up close to the ancient traditions of the Chanvriers (the French word for "hemp" is *chanvre*), who continue to model the harvesting of hemp and the traditional method of submerging the stalks into the Loire River to accentuate the microbial retting process and prepare the fiber for historic rope making systems, which are primarily still being done for cultural heritage preservation purposes.

Hemp grows fast and gets tall—fifteen feet of linear growth is possible with some varietals. This yield, in combination with the strength of its fiber, means hemp has the potential for developing innovative agro-ecological systems within our natural fiber and food systems. Hemp also outcompetes weeds and can be highly immune to pests and disease. The crop is relatively frost-tolerant as a seedling, thus lowering the risk when planting at higher elevations. As with flax, which also has a

North Carolina Hemp Farming

by Tyler Jenkins, community organizer

The first organic field trial of industrial hemp was planted in northeastern North Carolina in early June 2017. Prior to the project's beginning, we were tasked with three things: sourcing certified industrial hemp seed, getting a license from the state, and finding land for a trial site. Seed was acquired from Italy, and close working relationships with an international seed supplier, the North Carolina Department of Agriculture, and the DEA were leveraged to ensure delivery. Multiple legislative and market realities delayed timely delivery of the seed; however, once it arrived we were able to plant in less than seventy-two hours. Our partners included the Northern California Fibershed, Bountiful Backyards (an edible landscaping company), Earthwise Organics (organic systems guru in the southeastern United States), and Homegrown Agriculture (a bold and diversified farm in Eastern North Carolina).

We observed the hemp field trial site throughout the summer and documented a variety of findings on the eight varieties. We seeded at a very low per-acre rate—around ten pounds per acre—so that we could observe differences in each variety within the quarter-acre plot. Our part of North Carolina has clay soils, with the exception of sandy loams in the alluvial floodplains. Even with the late planting, the crop grew to an average height of four feet over the course of two months. We had an unusually hot and dry summer, with a few downpours late in June and one in early July, and the hemp crop thrived in the dry conditions. One interesting insight was the number of native bees we saw attracted to the flowers. We also observed insects we had not seen before in our fields. We harvested plants in August and were able to harvest fifty pounds of plant fiber to process further for textile development. We harvested flowers and leaves for a local craft brewery that made use of the terpenes for the creation of a hemp-flavored beer.

Organizing the trial allowed us to not only study the characteristics of the eight different varietals but also understand the possibilities and barriers for each stakeholder along the value chain. The farmers were able to assess the time and risks involved in acquiring a license and preparing land for a successful trial. We were able to take the hemp stalks to local fiber makers who have existing market segments in the fiber community and were excited about integrating hemp into their products. In short, the trial allowed us to "pilot" two ends of a supply chain and prepare ourselves for a seed-to-market project the following year. Some goals we have for the next stage are to increase the visibility of the crop, our partners, and the local fiber system; to bring a small, modular processing unit to North Carolina that takes hemp from retted fiber to spun yarn; and to market a product utilizing a replicable and comprehensively carbon-negative value chain to do so.

hundred-day growing season, producers could design rotations with other cash crops to maximize their on-farm returns, and these rotations could be designed to include the integration of livestock. By integrating sheep or goats into the farming system with hemp, producers could enhance nutrient cycling in their fields, clear remaining plant residues

The mechanical decortication machine developed by the One Acre Exchange. *Photo by Max Holden.*

before seeding the next crop without the need for herbicides, and potentially reduce or eliminate tillage. Including sheep and goats into the cropland rotation design provides another opportunity to stack an agricultural grazing-based business into the economic model and provides a potential opportunity to support non-landowning contract grazing operations.

Hemp is known to increase wheat yields; some research has shown up to 20 percent higher yields for wheat when it is planted in rotation with hemp. In the United States in 2015, farmers grew fifty-six million acres of wheat, most of it in the winter months, which leaves the possibility open for rotating with a summer hemp crop. If we were to conduct a thought exercise focused on assessing what it would look like to add hemp into the wheat cycle within the United States, we could assume yields of an additional 112 million tons of hemp fiber flowing into our textile system (if we assume a yield of two tons of fiber per acre). If these same croplands were integrated with fiber-producing livestock (sheep and goats) to graze and clear the fields, cycle nutrients, and support the ability for the farmer to be able to plant the next crop using conservation

Outside of Durham, North Carolina, the One Acre Exchange project, now led by local resident Tyler Jenkins, began growing an acre of hemp to evaluate the agronomic needs of the crop along with a focus on developing as many relationships with local makers as possible—including weavers, brewers, hand-spinners, and more. *Photo by Anna Carson Dewitt.*

and/or no-till strategies, this would support known soil health-enhancing practices while adding an additional 950,000 tons of millable protein fiber based on low stocking rates of seven wool-producing animals to the acre. (This amount of wool assumes 50 percent of the wool would be lost in processing, a normal loss for industrial milling processes.)

The Increasing Demand for Hemp

It could be fairly easy to assume that this integrated agricultural system would healthfully yield wool, hemp, wheat, lamb, and dairy and would generate an additional 112 million tons of natural fiber per year in this country alone *without converting any additional land to agriculture.* That could equate to approximately enough fiber to produce twenty-three *billion* one-pound garments (assuming weight losses to the fiber throughout the milling process). Nationally, we don't consume that many garments per year, which in turn means that we would have ample room within this scenario to reduce the acreage planted with hemp significantly and still produce enough natural fiber to clothe everyone in the country with compostable nontoxic garments.

Colorado Hemp Processing

by Arnie Valdez, farmer

The Rezolana Institute, also known as Rezolana Farm, is a homestead-focused research and demonstration project based in San Luis, in southern Colorado, that is actively engaged in integrated crop and animal impact farming techniques. The farm grows heirloom crops such as Anasazi and bolita beans, sixty-day corn, and "cottage-scale hemp" in rotation with multispecies cover crops that are grazed down by the resident goats at the end of the growing season. The Rezolana Farm is part of a communally managed watershed that is framed by the Sangre de Cristo, La Garita, and Conejos-Brazos mountain ranges. The snowmelt-generated water drains through hand-dug *acequia* (ancient trench) systems that move water through each farm and ranch within the community, as our limited water supply must be shared. Most of the farmwork is conducted with vintage farming equipment and hand labor for weeding and harvesting. The farm is gradually acquiring infrastructure such as season-extender hoop houses, water conveyance systems, composting systems, and hand-operated equipment for hemp processing.

Severe drought in the Southwest continues to be a formidable experience for the farming and ranching community in the region. In accord with our long-term soil organic carbon building goals, and in light of the drought, we have focused our efforts on supporting soil building on the farm's arable lands, including the use of compost applications, the integration of livestock, and a polyculture planting that utilizes the least possible quantity of water. The blend includes Rezolana Institute hemp seed, Arivica peas, Terranova radish, Purple Top turnip, VNS buckwheat, and Monida oats. Hemp has been harvested by hand on the one-and-a-half-acre field trial site and used for demonstration purposes at a hemp fiber and woody hurd processing class that we offered at the institute. We use hand-powered machines to separate the hemp fiber from the woody hurd. We've made open-source plans available online so that others can make their own hemp processing machines as we have. The original hand-powered fiber and woody hurd processing machine design originated from the work of Growing Warriors—a Kentucky-based veteran farming nonprofit.

In addition to the fiber that we produced on our farm, we have also used the hurd for manufacturing earthen blocks composed of soil, stabilizer, and hemp shiv. Rezolana Institute presented the results of our block experiments in a paper and presentation for the 2017 Earth USA conference held in Santa Fe in late September.

Unfortunately, these types of thought exercises are not led by fashion brands or wearers. Yet. We have little to no demand-side pull-through strategies that would engage farmers at this scale. And while farmers respond to markets and are very capable of making changes and adopting new practices to improve their personal economy, they need an indication that the market will support them. On the supply side, we also need to be able to provide technical support to land managers as they make the transition to integrative soil building and Climate Beneficial

The town of San Luis in southern Colorado produces sixty-day corn, heirloom beans, and a range of vegetables and is now also known for its emerging hemp agriculture projects. *Photo by Rio de la Vista.*

Hemp seed planting in San Luis, Colorado, in 2014; the first crop was planted after new laws opened the doors for hemp agriculture. *Photo by Michael Keefe.*

systems. Increased connectivity and subsequent agreement making between fashion brands and agricultural land managers could lead to an increase in diversified farming systems and potentially to early-stage funding that would help jettison new fiber and food projects into being.

When attempts to restore hemp farming to the landscape began in 2014, there was no available seed stock specifically bred for the climates where the plant was being reintroduced other than from feral varieties that had survived in a few small ditches and roadways. There was very little understanding of what varietals would be the best match for our latitudes, and universities were hesitant to jump into their historic role as agricultural technical assistance providers due to fears that they would lose federal funding, even though the plant was technically legal to conduct research on. Some farmers were fearful to grow the crop for research due to concerns that federal-level DEA agents would confiscate their land and the USDA would put their crop insurance in jeopardy.

Each week, it seemed, a new and previously undiscovered hurdle seemed to reveal itself. For example, once retted stalks are stripped of

Hemp brick making processes at the Rezolana Institute. *Photo by Rezolana Institute.*

their fiber and the hemp is ready for textile processing, there is a specific type of machinery and skill that must be used to handle long bast fibers. There are currently no large-scale, high through-put, long-staple fiber processing centers in all of North America. This meant that to get the ball rolling on an economic model that would keep farmers engaged in growing hemp, we had to develop methods for processing the plant with fibers for which we did have processing equipment. Within the existing milling systems in the United States, the most decentralized and technically malleable system is designed for processing wool, and so the first inquiry we made was to determine if we could blend hemp with wool and alpaca and process the fibers through woolen combing and spinning machinery.

The steps required to prepare the hemp for the woolen mill required another layer of technical expertise, however. The first technology partner that we had an opportunity to work with brought forth a fairly high-throughput machine that could separate the fiber from the woody part of the crop, known as a *decortication system*. BastCore, a company based in Omaha, Nebraska, stepped in with decortication technology

Hemp fiber grown at the Rezolana Institute being processed on a wooden hemp break built by Arnold Valdez of the Rezolana Institute and designed by Michael Lewis and colleagues in Kentucky.

that offered a method to separate the fibers from the woody parts of the plant while protecting the fibers from excess abrasion, retaining the textile-grade quality of the fiber throughout the process. Once the stalks were decorticated, the next step in the process of getting hemp ready for the woolen milling system was to further remove the remaining lignin from the fiber so that the hemp was soft enough to be blended with protein fibers. Our effort to soften the hemp (known as de-gumming) has been an ongoing project to research methods that are ecologically focused, such as the use of enzymes, recycled water baths, compressed liquid carbon dioxide, and UV light treatments. Continued refinements in these technologies are occurring in an effort to find the most ecological solutions to soften this fiber at a reasonable scale.

Nettle and Other Natural Plant-Based Fibers

There are many other plant-based natural fibers outside the boundaries of what we've come to know through our own bioregional explorations and Fibershed research. Each of the fibers mentioned in this section deserves pages of words to more deeply explain the cultivars and

Milkweed as Fiber

There are seventy-three species of milkweed in North America, many of which are rare, threatened, and endangered. Milkweed is the primary host plant for monarch butterflies and provides food for the young monarch larvae while providing a protective *cardiac glycoside*—a natural toxin that makes the monarch butterfly unpalatable to most predators. Unfortunately, monarch populations have dropped to near species-extinction rates on the West Coast of North America. The butterflies' overwintering sites are monitored in California in areas that provide protected habitat from freezing temperatures, and in the winter of 2018 fewer than 30,000 butterflies were counted by citizen biologists, compared with more than 192,000 in 2017, more than 1 million in 1997, and at least 4.5 million in the 1980s. The last count marks a 97 percent decline in the population. The Xerces Society—a nonprofit focused on invertebrate conservation—is calling for swift action to restore the breeding and migratory habitat. Habitat loss has occurred from the use of pesticides and herbicides in both residential and agricultural communities, where water tables have also dropped due to human use, and thus the muddy ponds and damp ground that hydrate the monarchs have disappeared as well.

Monarch populations in the eastern part of North America have experienced long-standing decline, with a bump in 2018 that showed a 144 percent increase in the population. What to attribute the population increase to is still being analyzed—but we do know that there are inspiring efforts to enhance monarch habitat in the very agricultural landscapes that were once known to make milkweed eradication a common practice through the use of herbicides. Projects in agricultural communities in Vermont and Canada, for example, are growing over twenty-two hundred acres of milkweed for its fiber specifically. In an interview with Dr. Heather Darby of the University of Vermont, it was explained that the fiber of milkweed, known as *floss*, has been proven to have thermal efficiencies that exceed all synthetic materials as well as natural materials such as goose and duck down. The floss is hollow and also hydrophobic (meaning it resists water). Milkweed fiber can effectively be used and has been used as sleeping bag fill, jacket insulation, and boot insulation. There are innovative approaches to use the material as an insulator in electric cars and also to support cleanup of oil spills. All of these material culture uses can be obtained while simultaneously providing the necessary food for monarchs.

Kimberly Hagen wearing her favorite Mae West milkweed jacket at the University of Vermont campus. Jacket designed by Charlotte X. C. Sullivan and Alayna Rasile Digrindakis, with batting produced by Monarch Flyway. *Photo by Suzy Hodgson.*

processing histories, and their beauty alone would justify a deeper inquiry. Dogbane, ramie, New Zealand flax, kenaf, kapok, lotus, milkweed, sisal, pineapple, and coconut (to name a few) are all species that that generate textile-grade fibers.

Nettles (*Urtica dioica*) are particularly promising despite being closely identified with stinging barbs of their leaves and the painful rashes they cause. This lustrous bast fiber originates within the nettle plant's stalks, and the fabric it produces is similar to linen, though stronger (especially when it gets wet) and a bit shorter. Though finer than flax linen, nettle fiber is a bit stiffer, which makes it useful for clothing and other textiles that need more structure, such as a jacket. It can also be blended easily with other natural fibers. Nettle use in textiles and for thread has been around at least since the Middle Ages, though there is now evidence its use extends back to the Bronze Age in Europe. The perennial plant was also cultivated for the medicinal and food properties of its barbs (as they lose their sting when cooked). Over time, nettle fabric was replaced by imported silk, cotton, and other fabrics. Recently, though, it has enjoyed a resurgence of interest thanks to new spinning technology, plant cross-breeding, and a concern about the toll of conventional cotton.

Nettle water retting. *Photo by Mari Stuart.*

The preparation of nettle fiber is similar to that for flax and other bast fiber species. There are a multitude of ways to accomplish the softening of the fibers. One method includes processes discussed earlier for hemp processing—the stalks are harvested and then laid on the ground to be aided in their deterioration by microbes in the soil, or they are soaked in water first for a period of time to further enhance the rate at which the lignin and pectin that makes the fiber stiff are broken down. The fiber can then be separated from the stalk and further mechanically softened through what is known as breaking, scutching, and hackling. Once those steps have taken place,

Nettle harvesting. *Photo by Mari Stuart.*

the fiber can then be spun into yarn. An important advantage of nettles is the ease and abundance with which they grow, utilizing both seeds and underground rhizomes to propagate. Nettle species can be found all over the world in many different types of habitat and soils, though the plant prefers rich, humic soil, which makes it an ideal indicator for archaeologists of former human settlements. There's more to love about nettle: Its stinging leaves discourage consumption by wildlife, and it is highly competitive with weeds. Nettle also produces a natural dye ranging from yellow to yellow-green. Its remarkable adaptability and utility was even recognized by Victor Hugo, who wrote in his famous book *Les Misérables*, "What is required for the nettle? A little soil, no care, no culture. . . . With the exercise of a little care, the nettle could be made useful."

Combing and spinning nettle.
Photo by Mari Stuart.

For a discussion on the history of nettle fiber production, I recommend *Through the Eye of a Needle* by English writer John-Paul Flintoff. The book documents the arc of nettle fiber usage within Europe. His investigation of nettle production became a hands-on personal mission whereby he attempted to make his own nettle garments, and in the process he came to appreciate the quality of the length of the fiber. He speaks to "discovering the virtues of the humble nettle," describing how we've come to know and appreciate the medicinal and local food value of nettle leaf, from teas to slow food soups, and yet we've forgotten that we can also make our own clothes from the remaining stalks. The book documents that this effort to grow endemic European materials and process these fibers into clothing has precedents going back as recently as the early twentieth century. There were historic clothing shortages in Germany at the time of World War One, due to sanctions imposed by Great Britain. The German military lost access to Britain's cotton empire, and in the process the German people began rediscovering nettle and began growing it at scale, even using it for their military uniforms. This recent history brought forth the realization that there were long-staple bast fibers abundantly available, requiring no external irrigation and relatively simple oversight during the growing season.

The Future of Plant-Based Fibers

Plant-based fibers and dyes have been sustaining human life since we necessitated a second skin for our survival, and they are the materials that will see us safely into the future (if we enable that to occur). The current status quo textile economy has put into question how much longer we can continue to provide an expanding human population with virgin raw materials from depleted land bases. There are, however, key mechanisms for transforming the current model that will allow us the greatest chance at continuing to bring forth beautiful, health-providing, and inspiring textiles well into the future.

Plant fibers have been shown to be able to endure recycling systems that turn old garments into new, allowing a plant-based garment to be renewed at minimum three and up to

> This recycling capability is an important facet of the overall road map for bringing natural fibers into the future.

five times in some cases. This recycling capability is an important facet of the overall road map for bringing natural fibers into the future.

Slowing down consumption is an equally important facet of the future—purchasing quality over quantity and creating a culture that treats garments like heirlooms is critical. With these important strategies at work, we can then turn our attention toward the question of how we will grow plant-based fibers and dyes in the future. We've explored the critical importance of restoring our soil carbon pool as a water saving and climate change amelioration strategy earlier in this book—efforts that have yet to be fully deployed. And we have yet to discover all the opportunities that await us as we begin to recover our soil's historic carbon content. When we combine these soil recovery efforts with creative approaches for stacking functions within our agricultural landscapes—integrating our food, dye, flora, and fuel systems—there is a nexus of efforts where the possibility of a healthy and truly regenerative fiber-and-dye future starts to reveal itself.

Designing integrated landscapes means that our fiber-and-dye systems could function in whole or in part markedly differently from today's monocropped fields and high-input production methods. There are many region-specific agricultural solutions awaiting our creative attention. In one of many diversified fiber-food-dye visions of the future, I can foresee within the temperate climate I live the possibility that plant fibers could be alley cropped within in an agroforestry system. Examples of tree species and their multiplicity of offerings include:

BLACK OAK, which yield nutrient-dense acorn meal and rich tannins that help bind natural dyes to fibers.
MULBERRY TREES, which produce superfoods for humans and leaf forage for silkworms.
BLACK WALNUT TREES, which produce human-palatable proteins and color-fast deep brown dyes.
OLIVE TREES, which produce oils rich in vitamin E, vitamin K, and monounsaturated fats, as well as leaves that yield pink and gray dye palettes.

Each of these tree species can provide cover for an understory of many species, including dogbane, flax, hemp, or cotton that can be seasonally rotated with heirloom grains and polycultures of vegetables. Each of these understory crops could be seasonally pulse grazed by hearty heritage-breed sheep, goats, and cows to help cycle nutrients and provide additional fiber and food. The bast fiber species—particularly hemp and flax—offer rich fatty acids and medicine that can be harvested as a main focus of the crop's human use. Fiber could be a by-product in some if not all cases of a food and medicine harvest regime.

In the land bordering these integrated agroforestry systems, I can foresee hedgerows planted for migrating bird habitat and food sources, as well as pollinator forage; natural dye harvesting and pruning processes could help maintain these plantings—further blurring the interface between what we think of as wild and domestic. Whatever the future of our fiber, dye, food, fuel, and flora systems might be, they will surely have the best chance of meeting their intended ecological outcomes when the design is led by those who know and love their connection to a particular place and have equity and access in and to the land to apply their stewardship know-how as part of a consistent and lifelong practice.

Danielle Svehla designed and knit this sweater made of undyed alpaca yarns from Renaissance Ridge Farm in Mt. Aukum, California. Danielle worked on the piece while completing her PhD from UC Berkeley; her research focused on the impacts of climate change to mountain regions and their forested landscapes.

CHAPTER SIX

Implementing the Vision with Animal Fibers and Mills

*T*here are many animals that generate their own naturally warm coats. Over the millennia they have shared their fiber with humans. Originally this occurred through shedding their coats on the thorns of berry bushes and on the bark of trees, which provided humans access to tufts that could be collected for felting, spinning, weaving, and knitting. This activity occurred long before we began domesticating our animal cousins and providing them with annual haircuts. Humans have long relied on animals for food, and in the process of hunting and curing animal proteins we have in turn made use of the whole animal, processing the skins and hides. These materials were historically the foundation for human shelter, in the form of both clothing and homes.

In this chapter I will look primarily at sheep's wool (mainly because this fiber has become the predominant protein fiber in my region) and its role in defining our regional fiber economy. One key facet in the development of a regional fiber-and-dye system is to first consider what is already abundant and available and how to enhance the access points between farmers who grow and wearers who wear. It's wonderful to diversify and build out our regional fiber-and-dye systems into new textures and colors and also important to stimulate community building to focus on the agrarians who are already hard at work producing these useful materials.

Wool: The Key Ingredient to Local Clothing

Wool has been co-evolving with the forces of nature and human behavior for many millennia. One of the best-known wool producers, the sheep (*Ovis aries*), was domesticated to live alongside humans between eight thousand and ten thousand years ago, beginning in Mesopotamia and Iran. At the time they were more hairy than woolly, so they were valued primarily as a source of milk and meat, whereas the cultivation of sheep for wool took place over the next two thousand years. The first archaeological evidence of woolen fleece dates to roughly four thousand years ago with the appearance of specialized textile-making tools. One of the earliest pieces of woolen cloth discovered is more than three thousand years old, perfectly preserved in a Danish peat bog. Over the centuries, domesticated sheep and their shepherds spread into many microclimates across the planet, adapting and adjusting to a wide variety of temperature ranges, precipitation patterns, and terrains. During my one-year wardrobe challenge, wool was the most accessible fiber available for garments made within our bioregion and remains so to this day.

The natural properties of wool make it an exceptional material to create our clothing. Its properties are unique; wool absorbs water vapor both from the air and from the wearer's perspiration through the porous coating that covers its scales. The fiber can take in 30 percent of its weight in water vapor without feeling damp or clammy. Wool is also a well-known temperature regulator—energy is released when water vapor enters the fiber, and this offers warmth to both the sheep and the wool wearer. A kilogram (2.2 pounds) of wool can produce as much energy as the human metabolism generates within one hour. Conversely, when wool enters a dry and warm environment, the fiber produces a cooling effect as the fiber re-releases moisture. The flame resistance of wool is also remarkable—it has the highest ignition threshold of any natural fiber (as high as 600°C—over 1,100°F) and is self-extinguishing. Wool is biodegradable—the fiber can decompose within months depending on how it is composted or returned to the soil. Wool is known for its utility, and for many people it is their first choice for high-performance activity, whether that activity is marathon running or herding sheep. The fiber is malleable, durable, and comfortable, making it useful for all seasons and all types of activities.

The diversity of wool types is analogous to the variety of wood products you could use for the construction of a house. A carpenter knows that certain wood types are better for beams or planks than others, and not just any wood can be used for any application; the material has to suit the lengthy occupation of the house and the functional demands it is designed to serve. Wool is similar to wood in this way; there are many types of wool from many breeds, including goats, across the world, and they serve many different purposes, though I am focusing on sheep in this chapter.

Why Breed Matters

As wearers of wool clothing, many of us have only begun to skim the surface of what different breeds of sheep produce and what this variety of material could mean for creating highly specified, durable, functional, and heirloom-quality garments and utilitarian goods. While exploring our fibershed in the quest to make garments for the wardrobe challenge, I discovered a wide range of wool types and sheep breeds. There was, and still remains, a group of flock owners in the area, primarily women ranchers, who through the practice of raising sheep over the years and producing their own yarn have come to deeply understand the varied attributes of the wool their animals grow. Each person in this small and dedicated group has focused their life's work on the production of high-quality fiber through artful breeding practices. I was eager to learn as much as possible from them. I spent afternoons in the hot inland irrigated green pastures of Orland, California, learning about the luminescent qualities of Cormo wool. Several hours on the road farther west, we found ourselves in the pastures adjacent to Tomales Bay, enjoying the rugged texture and colorful fleeces produced by heritage Churro sheep.

After digging into historic sheep breeding reference documents in combination with some hands-on experimentation, it became quite evident to me that wool produced from breed-specific flocks was destined for specific outcomes and to try anything outside of this well-established canon of use was ill considered. An abundant breed of sheep in our region, known as the Corriedale, produces a type of wool that has proven very useful for sock making. Durable and soft, but not too soft, the Corriedale fiber holds up well to abrasion. Socks made from Corriedale wool last twice as long, at a minimum, as those socks made from softer and finer wools. The natural colors of the Jacob fleeces made beautiful, natural multitoned garments and also socks, just soft enough to be worn next to the skin. No dye was required to create

> Socks made from Corriedale wool last twice as long, at a minimum, as those socks made from softer and finer wools.

Icelandic sheep raised by Sophie Sheppard. *Photo by Sarah Lillegard.*

colorful gray-toned knit pieces. Romney and Churro wool can be felted into durable brown, gray, and black carrying cases and wide-brimmed hats; and yarns made from pearly white Cormo wool can be used for a range of next-to-skin-quality garments, including those worn around sensitive areas like the neck and wrists.

The texture of these wool types and the various ways in which the wool is spun into yarn offer unique physical responses. These breed-specific yarns bring another layer of individual responses when dipped into the natural dye vat. A merino wool takes the dye color with a kind of matte finish, while a Wensleydale wool will reflect the dye color in an opalescent manner, similar to a gloss finish. These dye uptake variations in turn lay the foundation for a completely new set of permutations of color and textural forms previously unknown to many of us who have historically considered wool as a homogeneous fiber. The diverse color and texture explorations taking place in our fibershed, and the abundant human and natural resources that they represent, make it hard to imagine that we could ever reach the limits of wool fiber possibilities within a single lifetime.

When we take our lead from the land, many textile possibilities begin to present themselves. All these color and form possibilities are generated from a simple foundation: Sheep grazing on grass happens to be well suited for the coastal mountain ranges of California as well as the slopes of the eastern foothills in the Sierra Nevada range. Sheep also do well in our state's Central Valley, including when they are grazed in the vineyards and orchards. Our fibershed community raises the following sheep breeds: Shetland, Romney, Corriedale, Merino, Rambouillet, Romeldale, California Variegated Mutant, Suffolk, Churro, Jacob, Icelandic, Bluefaced Leicester, Cheviot, Ouessant, Targhee, Columbia, Dorper (that do not produce quality fiber), Perendale, Southdown Babydoll, Wensleydale, Cormo, and a host of hybrids. In our area there is also a good representation of Huacaya and Suri alpaca as well as mohair goats, llamas, Angora rabbits, and some beautiful guanacos (camelids native to South America). The range of breeds is possible because of both the diversity of climatic zones in our region and the inquisitive and persistent flock owners who seek diversity and preservation of heirloom breeds as part of their agrarian missions. This diversity is not unique to our state and can be replicated in many other parts of the world.

When a community attempts to answer questions regarding what type of land-based economic development projects are possible, it is key to first become a student of what the land can produce or already is generously and abundantly producing. Answering the question about what is physically possible from an animal fiber textile perspective has to start with assessing pastures, crop residue opportunities for grazing, and the forage in rangelands. The ecology of place—the rain, fog, wind, forage, soil, and seasons—defines the quality and quantity of wool in any region. For instance, on the coast of California, especially in the regions of Marin, Sonoma, Mendocino, and into Humboldt Counties, you can see that the large flocks in these ecological zones tend to produce a significant quantity of coarser high-micron-count wool. This wool is of a quality well suited for some types of outerwear, a hardy pair of socks, bedding, housewares, and other durable goods. Many of the sheep on the coast are exposed to a fairly continuous influx of coastal fog. This moisture keeps our coastal temperate forests in health and vitality and also defines the parameters of what sheep breeds will do well in these conditions.

Corriedale, Romeldale, Wensleydale, Suffolk, and Ouessant (to name a few) are breeds that have a slightly coarser fleece and larger-diameter (micron) fibers, and this allows them to thrive in coastal

Sheep and Vineyards

There's been a renewed effort among some of California's vineyards and winemakers to implement soil-building and restoring practices. These include composting leftover grape skins and stems, which are then added to the soil to boost carbon stocks; planting cover crops between the rows of grapevines in order to protect the soil from erosion; attracting beneficial insects with the type of cover crops they plant; and eliminating fossil-fuel-based fertilizers. But by far one of the most intriguing has been the introduction of sheep grazing. The idea is that integrating sheep into cover crop/vineyard systems utilizing agroecological practices generates positive benefits for improving soil health, nutrient cycling, and carbon sequestration. However, very little scientific data had been collected and analyzed about these purported benefits until recently, when Kelsey Brewer and primary investigator Dr. Amelie Gaudin of the department of plant sciences at the University of California–Davis decided to conduct a monitoring study focused on these sheep-grazed perennial cropland systems.

Gaudin's team monitored and compared vineyards sites that had been grazed by sheep and adjacent sites that had been mowed mechanically (both practices are utilized to manage biomass between the vines). In the grazed plots, the soil carbon stocks trended higher than the ungrazed ones, including eleven thousand to fifteen thousand additional pounds of carbon per acre in the grazed plots (the paired plots contained similar soil types, slopes, vine/rootstock varietals, and irrigation treatment). The total amount of bacteria and fungi and the diversity of these microbes were higher at shallow soil depths in the grazed plots (this is called the *labile layer*, where a great deal of carbon cycling takes place), including the presence of certain beneficial microbes such as *arbuscular mycorrhizal fungi*. The amounts of *saprophytic fungi*, which are important for carrying out decomposition of plant materials, were notably higher in the grazed samples. The grazed plots also had higher levels of plant-available phosphorus, a critical mineral for plant health. Nitrogen pools were higher in the grazed plots, which was likely associated with the urine from the sheep.

The analysis confirmed what advocates of integrated sheep-vineyard systems (ISVS) had been touting: "Our results support that ISVS has substantial potential to increase soil C [carbon] storage and improve important ecosystem synergies such as microbial functioning and biogeochemical cycling," said Brewer.[1]

Robert and Jaime Irwin of Kaos Sheep Outfit move their flock into the vineyards in Mendocino County to graze down the vegetation between the vines.

landscapes. Moisture passes through their wool coats more easily than it would move through the more densely coated fleeces of the fine-fiber sheep, such as the Rambouillet and Merino. As you travel inland into our hotter and drier valleys and into the foothill regions, sheep like the Merino and the Rambouillet, which have denser and finer fleeces, thrive in these drier climates. By understanding microclimates and the biological diversity that is generated from various climatic zones, a picture of our textile-making possibilities begin to unfold.

Alpaca, Yaks, Goats, Rabbits, Silkworms, and More

There are many other types of animals with very fine fibers that are being raised in our fibershed and other fibershed communities, including Angora and Cashmere goats, Angora rabbits, Suri and Huacaya alpacas, guanacos, camels, yak, and musk oxen, all of which contribute very unique, soft textures and dynamically complex colors to the textiles produced from their fibers. Cashmere goats, for example, produce a fine-count fiber and physically do well at higher elevations, as the cool temperatures instigate the need for the animals to produce larger quantities of super-soft downy fiber.

Yak Fiber

Yaks produce a very soft and fine fiber that is appreciated for its largely black and brown colors. The Chokurei yak project in Saguache County, Colorado, at seventy-nine hundred feet in elevation on a plateau below the Sangre de Christo mountain range, produces small quantities of naturally black yak fiber for use by hand-spinners. Another yak fiber project that has come into full development is centered in the birthplace of the yak in the Tibetan Plateau. Known as Norlha, it was founded by a group of herders, spinners, weavers, and knitters dedicated to producing exquisite and world-class textiles all within their regional fibershed. One of Fibershed's designers, Danielle Svehla, was invited to the Norlha project to support the artisans in sewing and knitting pattern development, demonstrating how regionally focused fiber systems can support and enhance one another's efforts at refining the terroir of their own textile development.

Both yaks and goats produce very small amounts of millable fiber per animal—approximately four ounces of cashmere-grade fiber per year from one goat. The animals must be carefully combed to retrieve this soft, downy layer of fiber. Yak fiber is similar, in that the downy undercoat that is softest and highest quality must be hand-combed from the animal each year. These are delicate fibers, and it is important to consider that these materials necessitate high price points.

Alpaca

Suri and Huacaya alpaca are domesticated South American camelids, listed formally as livestock by the United States Department of Agriculture in 2008. Five years later alpaca production in the United States totaled approximately 178,000 animals. The Huacaya alpaca make up the vast majority of that number and produce a soft and slightly curly fiber, whereas the Suri produce a longer, silkier fiber with considerable drape. Alpaca fiber has an air bubble within it that makes it an incredibly insulating fiber.

Alpacas are sheared annually. The raw fiber does not contain lanolin like sheep's wool; it thus requires a lighter level of washing and fiber preparation. Alpaca fiber, unlike wool, has no scales and for that reason does not affect human skin the way that wool does. Thus, adding alpaca to wool can improve the hand of a textile. Alpaca on its own accord generates exceptionally soft and beautifully draping textiles, and many alpaca producers prefer to produce yarns with only alpaca fiber to retain the integrity and inherent properties of this fiber in the value-added product. However, alpaca can also be blended with sheep's wool (as can any plant or protein fiber mentioned), though it tends to offer specific properties to wool that improve the intrinsic qualities of a yarn—wool has crimp and alpaca does not. Adding wool and alpaca into a single yarn can create more bounce to a finished textile and more ability for a garment to return to its original shape after it's been stretched.

While the United States does not have a long history of alpaca fiber processing, there are individuals working to make milling accessible to producers. An innovative processing model in New England, known as the New England Alpaca Fiber Pool (NEAFP), has provided alpaca growers across the country a means to have their flock's fiber turned into all manner of beautiful finished goods including gloves, hats, scarves, coats, boot inserts, knitting yarn, and afghan blankets.

Angora

Angora goats and rabbits are small herbivores that we have seen raised on scales from a tiny urban homestead to a thousand-acre ranch. Angora goats produce *mohair* fiber at a rate of approximately three-fourths of an inch per month. The soft fiber locks of mohair have an incredible sheen and smoothness. Named after Ankara, the historic Turkish province where they have been bred and have thrived for many centuries, Angora

Fibershed member Marnie Jackson demonstrates hand-shearing an Angora rabbit at the Fibershed Wool Symposium.

goats were not exported from the region until 1849. The Angora goats that we've worked with in our fibershed have a consistently playful and yet docile approach to life. It is heartening to see little curly-haired goats frolicking in a pasture with their friends.

Angora rabbits are the only animal that produce angora fiber (Angora goats produce mohair), one of the warmest known fibers within our menu of options, and can be raised in your backyard. These highly soft and fluffy mammals produce small amounts of fiber, yet its warmth-giving properties go a long way if angora is blended with other more abundant materials like wool. There are German and English Angora rabbits, and both need to be sheared every ninety days to four months so that their fiber does not get too matted. Some hand-spinners will sit their rabbits on their laps, shearing or combing the animals while moving the fiber onto their spinning wheels. Both Angora rabbits and goats produce fibers that complement sheep-based wool yarns very nicely, adding sheen, warmth, a smooth texture, and weight. If you've ever worn a 100 percent angora sweater, you've quickly learned that some angora stays behind on you every time you take your sweater off. These fibers will shed, which is another reason to blend them with other fibers.

Silk

Silk is another protein fiber that dates back to the Neolithic era, derived from the cocoons of the silkworm. Raising silkworms for silk production, or *sericulture*, began in China between five thousand and seven thousand years ago. Silk produces a fiber that is known to be one of the best materials for skiers, backpackers, and endurance athletes who perform in cold climates because it wicks moisture and dries quickly. One hundred percent silk and wool-silk blends are favored materials for base layers and long underwear. Today there are many different indigenous varieties of silkworms raised in more than twenty countries, all of which feed on mulberry leaves.

Once mature, the silkworm spins a cocoon made of long filaments. In typical silk production the worms and their cocoons are boiled. In traditional communities the worm is then eaten as a source of protein, and the silk is harvested for the creation of yarn. The long filaments that come from the cocoon are extremely strong and can measure from five hundred to fifteen hundred yards in length, which is quite substantial given the source. There are silk creation processes that generate a fabric known as peace silk; instead of boiling the worms once their job is done, these methods allow silk moths to escape from their cocoons to live out their lives (which last a week or more), mating and laying eggs before they die. However, it is important to consider the role of the silkworm as a food source for communities that often have few other sources of protein. As of yet we have not seen a scaled silk project arise in our region or within other affiliate Fibershed organizations, but there has been a great deal of conversation about what that future project might look like.

The Importance of Mills

Processing mills are the most significant infrastructure investment our communities need in order to decentralize the textile industry and reinvigorate regional economies. However, the overall situation with centralized milling and large-scale commodity farming methods presents both a challenge and an opportunity to the fibershed model. Transforming fiber into locally made textiles involves various milling processes unique to each source, including fiber opening, blending, combing, and spinning.

Imagine living in a community where no one had a kitchen, and the closest refrigerator, stove, oven, and cooking pots and pans were all located overseas. There might be one person who still knew the art of cooking and had a few tools to make rustic meals, but it was never enough to feed everyone. Almost everyone would order their food, and it would arrive from overseas in boxes with no ingredients list. In the case of fiber, mill systems are the "kitchens" within the system, and the one person who still knows how to "cook" can be compared to a hand-spinner, weaver, or expert knitter. We'd never accept a food system with this level of infrastructure dysfunction, and yet we are all too willing to accept this level of archaic and unskilled design with our fiber system. We are currently desperately in need of more "kitchens," which means we need milling systems in every fibershed.

During my wardrobe challenge year, I learned through physical day-to-day experience the nuances of how milling wool and protein fibers impacts the body of the wearer. Imagine the experience of using thick sweater-weight yarns for the construction of most of your everyday items of clothing; in many cases, you'd lose the freedom of movement. One detail became clear: The gauge (thickness) of the yarn is instrumental in allowing you to move easily through day-to-day activities. Our locally milled and handspun yarns were great for sweaters and accessories but not for an item of clothing you would wear as a base layer or pants. Milling processes are the linchpin for defining the type of garment that can be constructed from a local fiber pool. If a local mill could offer yarn fine enough for more advanced knitting or weaving machines, our locally produced fiber could be turned into usable, everyday garments that could cover the entire human body. The question is how small and how regional can a local mill be and still perform this function at reasonable prices?

> **If a local mill could offer yarn fine enough for more advanced knitting or weaving machines, our locally produced fiber could be turned into usable, everyday garments that could cover the entire human body.**

In 2013, to address the critical questions of fiber supply and infrastructure capacity in California, we launched the Wool Mill Vision Project with the goal of gathering data on wool-producing enterprises across the state. The project was designed to measure the quality and quantity of California's wool in order to determine what end products it might produce and what kind of infrastructure it would need to do so. By conducting a supply analysis, we aimed to shed light on the quality and quantity of our existing fiber resources, and by designing a value-added infrastructure vision that matched the supply, we intended to illuminate what a regional soil-to-skin and soil-to-durable-good system would cost. This included how many people the mill would employ and what kind of products we could offer to the community at large. Our hope was that the project would facilitate the start of new businesses and be the catalyst for inspiring deeper relationships among textile brands, designers, artisans, and producers, while creating jobs in the region.

Our aim was also to address some of the big-picture challenges, including scaling up the production and lowering the price of fiber products to consumers. We believed that as shared production facilities and network coordination improved, small fibersheds could link together in pan-regional networks to share resources, improve supply, and reduce costs. While California produces over three million pounds of wool per year, the milling capacity in the northern portion of the state had been reduced to approximately ten thousand pounds annually. This has created a significant bottleneck, depriving wool producers access to an increasing demand from wearers and restricting access among textile designers for high-quality, regionally sourced wool.

Wool is a bit of a "low-hanging fruit" within our local fiber system. Many small- and medium-sized flock owners—who own sheep primarily for grazing management, dairy, and lamb production—have been unsure what to do with their wool after their animals have been sheared, making this an abundant but undervalued resource. When managed properly, sheep can provide a wide range of ecological benefits, from weed control to fire prevention and soil building. They are susceptible to predation, especially lambs, which adds to the financial challenges of raising sheep. Even in an era of hotter and drier conditions under climate change, sheep have the capacity to be grazed effectively on the state's range and croplands.

At the time we initiated our study, there were numerous conversations taking place with farmers and ranchers in our community about how wool had been viewed for more than fifty years as an agricultural by-product due to its low financial return. Lamb was the principal source of the income for many of the full-time sheep ranches and farms. But the wool market was beginning to change as a result of the recent surge of interest in the localization of food economies and the benefits of progressive land management. This interest has been driven largely by consumers as they sought healthy, high-quality products and supply-chain transparency around where their clothing and food originated, how it was processed before reaching them, and how animals were treated on the farm or ranch. We were well aware that wool could have an improved economic future, demonstrated by ongoing success of renowned enterprises such as Harris Tweed, Faribault Mill, and Pendleton Mills, just to name a few, each of which produces a diverse line of woolen garments.

How Much Wool Do We Have?

In seeking to understand our undervalued wool supply from both a qualitative and quantitative perspective, Fibershed conducted a comprehensive five-month study on California's annual production. One of our strategies included visiting a cross section of small- and medium-scale farms and ranches at shearing time to meet the sheep shearers and the ranchers. Together we'd fill out a short list of questions about flock size, breed, wool color, and the price that the flock owner received for their wool. We asked for two fingers' width of wool from the softest and coarsest sheep in the flock and were open to receiving wool samples from other individual sheep in the flock as well. I personally attended over thirty sheep shearings that year to collect wool samples and worked with the families and shearing teams to skirt the wool (remove short fibers and manure) while the flocks were being shorn. It was during these visits I saw for myself the skill and craft required to become an adept sheep shearer.

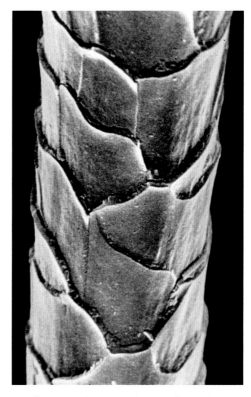

The sheep herds on these ranches were often too small to bale the wool into conventional square bales, which meant it often became impractical to transport the wool to a commodity market drop-off point for processing. Instead, many producers we encountered put their wool into large garbage bags and hoped someone would come to the ranch to buy it. Eventually, if the wool

An electron microscope image of a merino wool fiber. *Photo by CSIRO.*

did not find a buyer, it would end up stored in the barn, or used as compost in the orchard, or in a worst-case scenario tossed in a ditch or sent to the landfill. After spending time with the shearers during these information-gathering visits, we'd leave a packet of questionnaires with our organization's address on it and a pre-stamped envelope for them to take to future farm and ranch shearing sites. We offered them ten dollars for every set of wool samples and questionnaires we received back.

This outreach strategy allowed us to touch and feel numerous wool samples from all over the state grown by a group of small-scale producers who were unsure about what else they could do with this raw material.

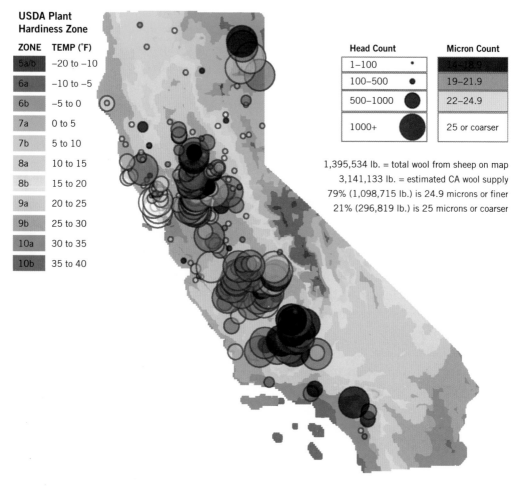

USDA Plant Hardiness Zone

ZONE	TEMP (°F)
5a/b	−20 to −10
6a	−10 to −5
6b	−5 to 0
7a	0 to 5
7b	5 to 10
8a	10 to 15
8b	15 to 20
9a	20 to 25
9b	25 to 30
10a	30 to 35
10b	35 to 40

Head Count		Micron Count	
1–100	·	14–18.9	
100–500	•	19–21.9	
500–1000	⬤	22–24.9	
1000+	⬤	25 or coarser	

1,395,534 lb. = total wool from sheep on map
3,141,133 lb. = estimated CA wool supply
79% (1,098,715 lb.) is 24.9 microns or finer
21% (296,819 lb.) is 25 microns or coarser

Graphically organized mapping of California sheep flocks. Larger circles denote larger flock sizes; darker circles denote lower-micron (finer-wool) flocks. Data sources: USDA Agricultural Census 2007; Roswell Wool Auction Data 2012; Fibershed Wool Survey 2013 (2012 sales numbers); Climate Zone Map, PRISM Climate Group, Oregon State University, created June 2, 2013. *Map by Amber Bieg, courtesy of Fibershed.*

The qualitative testing included assessing the micron count of the wool, which provided us with an average fiber diameter measurement and additional details such as the change in diameter over the length of the fiber and the standard deviation within the whole sample. We were able to send our wool samples to a former USDA employee who had access to the appropriate micron testing equipment. In exchange for receiving access to the wool samples, we provided the micron results to all of the farmers and ranchers who participated in our analysis. Thus with a fairly small financial investment on the part of Fibershed, we were able to start mapping the micron counts of California's wool supply.

By testing the coarsest and finest wool within a flock, we gained better understanding of a wide range of fiber texture. By correlating micron data with flock size and how many sheep there were of a specific breed on a ranch, we were able to make approximations in regard to total numbers of sheep and the quality of wool they were producing. To further our understanding we were provided micron core samples from the commodity market broker who aggregates wool from large ranches in our state. These core samples were taken from the center of large bales of wool that weigh anywhere from 450 to 500 pounds. We were provided the zip codes of sites where these core sample micron results originated and total poundage figures for each zip code. Once aggregated, we were able to map the quality of all the wool annually produced in California. This mapping process laid the foundation for us to understand what could be done with the wool on a region-to-region basis within California.

Fibershed inventoried 1.4 million pounds of wool, nearly half of the state's annual production. To our surprise, nearly 80 percent was determined to be fine enough for some form of garment production. This suggested an important opportunity for the development of local infrastructure. Counterbalancing this opportunity, however, was a steady, nationwide drop in demand for wool products over the decades, which one study attributed to rising consumer acceptance of clothing made from nylon, polyester, and acrylic. However, demand for high-quality wool fiber has risen sharply in recent years, led by many companies, including Smartwool, Duckworth, Icebreaker, and Rambler's Way.

According to our study, however, a significant challenge to California's wool producers is price. A variety of forces, including a glut of low-price wool from New Zealand and Australia over the years, have combined to keep market prices for wool low. The recent rise in demand for soft, high-quality wool has pushed up prices for the low-micron-count wool that is used in next-to-skin garments, but our state produces

John Sanchez shearing a sheep at Duckworth Farm.

a range of micron counts of wool, not just fine counts. Another challenge is a critical lack of any form of processing infrastructure; there are no regional mills that are able to open raw wool bales mechanically, nor at a functional scale to match the sheer quantity of the supply, washing, combing, and processing that fiber into yarn and knitted or woven goods. All large-scale mills in our state have closed since the heyday of wool production in the early twentieth century, leaving a crucial void.

Developing the Wool Mill Vision

Even though wool numbers will vary from year to year depending on a variety of factors (including fluctuating levels of precipitation, forage conditions, outbreaks of wildfire, disease, markets, and the goals of policy makers), what the wool supply analysis taught us was that we do have appropriate quality and quantity of fiber to generate head-to-toe clothing options for our fibershed community with the right infrastructure investments. A vision emerged from our work—one of heirloom, homegrown textile production derived from a thriving system of farms and ranches engaged in regenerating soil and ameliorating climate change.

We shared our wool supply analysis with engineers at Gaston College in North Carolina, former Kent Wool Mill employees, and equipment manufacturers from textile machine manufacturer Schlumberger and Italian spinning company FANI Pinter Group, all of whom made recommendations on what kind of equipment could process our wool into the highest-value goods. We reverse-engineered our favorite wool textiles and shared our findings with the engineering team, who helped us refine a factory-floor road map that would take greasy wool and produce three weights of naturally dyed wool knitted fabrics. Our design goal was—and continues to be—to create performance-quality fabrics that are 100 percent biodegradable and delivered to a range of cut-and-sew partners with a net negative carbon footprint.

To accomplish this goal, the team assessed the net quantity of atmospheric carbon we could sequester per hectare on sheep-grazed landscapes using our best understanding of appropriate carbon farming practices. We assessed a suite of renewable energy options to run the entirety of the mill, the living machine (a patented form of bioremediation of wastewater) requirements to recycle the water in the building, the operational schedule, labor requirements, and training schedules. We then assessed three viable locations to build the facility and determined land and building costs.

Resilience and the California Fires

In fall 2017 wildfires began burning throughout California, greatly impacting producers and their lands and livestock. Thousands of acres of pasture burned over the course of several weeks. Fibershed staff and producer community members provided support and communication to one another, checking in as the fires advanced. Many producers lost power and access to the internet for varying amounts of time or needed transportation help or safe places to take their animals. Our community worked swiftly to link people and animals to resources, including livestock haulers, evacuation notices, livestock hosts, and news during a time when uncertainty and weather variability ran high. A shift in the wind could change things in a moment.

Afterward as people began to look toward cleanup and recovery, Fibershed conducted a needs assessment to understand where we could be of help. We concluded that the greatest need was making funds quickly available to cover replacement hay for winter feed. With its pasture, barn, and winter feed burned, Joshua Farm Shetland Sheep had mouths to feed that couldn't wait for an insurance check to rebuild the barn. Selling sheep in the winter, right before lambing season, would be a significant loss for the farm, as well as for their land management come the spring. Using money we collected from vendors at that year's Wool and Fine Fiber Symposium, as well as cash donations from attendees, Fibershed contributed to over half the cost of Joshua Farm's winter feed, enabling their recovery without the farm incurring any further losses due to lost available feed or sold pregnant ewes.

We live in fire country. As a part of building a resilient regional textile system, we must work together as a community when disaster strikes and continue to reach out and support one another through these vital networks of producer communications and outreach.

We discovered we could produce a midweight wool knit textile that could be designed and sewn into a mid-layer top to keep a human warm under fairly cool conditions. This knit fabric could also be used for an athletic-type pant good for movement and activity. Both garments could be created with California's wool and were deemed to be the highest-performance design that could potentially be manufactured based on our raw material supply. Our wool was also of a quality well suited for heavier woven jacket- or shirt-style outerwear, wool coats, and pants that could be worn to keep warm and dry, and could be tailored to ensure mobility. The realization that these types of textiles could be made with our local resources was an important moment for our little research team. When we'd previously inquired about California wool with local designers, brands, fashion schools, or even some of the farmers and ranchers themselves, we'd been told that it probably wasn't good for much. We were prepared to assume that we'd be filling our mattresses with it (a perfectly wonderful use); we didn't know if our wool would meet the specifications for performance-level garments.

Next we met with wool textile mill equipment manufacturers to share the results of our supply analysis and textile deconstruction tests. We learned that our quantity and quality of wool justified the creation of what would be considered a "boutique mill" by global milling standards. This eighty-five-thousand-square-foot facility would begin by processing 1.3 million pounds of wool per year and would be technically large enough to process up to 5 million pounds a year when running at full capacity—a larger quantity than our state was currently producing. However, the mill design would provide room for growth and expansion of sheep numbers and wool production.

The mill was based on the premise that it would purchase "greasy wool" straight from the rancher at the point of shearing at a revalued higher price than what producers were currently receiving, providing the farmers and ranchers greater revenue streams and thus a renewed interest in focusing upon quality wool production, soil building, and climate change ameliorating land management. Based on the engineering specifications that the wool mill equipment manufacturers provided to us, our team developed a road map for how a California-based mill would function, from defining the necessary job descriptions to outlining the renewable energy needed to power the systems, determining the rainwater catchment and water recycling systems required to maintain a light groundwater footprint, defining possible locations for the mill's existence, and scoping out who would purchase a textile from it.

The Wool Mill Vision is a seventy-five-page document that outlines how to create a closed-loop manufacturing model with a clear and significant price tag of twenty-six million dollars. Again, this was considered a small mill by global standards, and in terms of our state's history, we had a track record of established and profitable working wool mills through the late nineteenth century. The idea of adding value to our region's wool has historic precedence; California once hosted *twelve* vertically integrated wool mills. However, our investment carried some level of risk. The cost of goods sold would in some cases be 20 percent higher than analogous versions made overseas, and the mill would produce upward of 640,000 yards of fabric per year, requiring a solid customer base.

> We learned that our quantity and quality of wool justified the creation of what would be considered a "boutique mill" by global milling standards.

It was challenging to overcome the modern cultural perception that importing our clothing and textile resources from outside our region and country is part of "progress." The idea of investing in physical systems of onshore manufacturing remains relevant only to past generations, and some of these sentiments were challenging to overcome within the investment community. Since then, the perceptions have begun to migrate in a new and positive direction, and we are beginning to see greater commitment to building the infrastructure for textile systems here in our home communities. From 2009 to 2014, for example, textile mill infrastructure investments have seen an 87 percent increase, while industrial production is up 7.6 percent for textile mills compared with 6.1 percent for all US manufacturing.[2]

For our fibershed at the time, twenty-six million dollars seemed high (as you can imagine), and yet this is the investment requirement to achieve naturally dyed biodegradable performance wear, carbon farmed and processed transparently within one region. We remain in a set of dynamic conversations about how to manifest this vision on the ground.

Regionally Scaled Infrastructure Is Critical

To compete within the conditions of modern liberal trade policies that remove quotas, lower or eradicate tariffs, and do not regulate for labor or environmental protection between countries, textile manufacturing has had to adapt and in most cases has gone the way of finding the lowest-common-denominator standards for labor and environmental protection. One key aspect of the industry's adaptation has been the centralization of milling centers, whereby increasingly larger-footprint mills have been moved to countries with lower labor costs and weak environmental protections. To this day the vast majority of the world's milling processes remain in China and Southeast Asia.

China has continued to surpass all other countries in textile exports. This movement and centralization of the industry has had economic and cultural consequences. Within the textile industry alone, the United States lost nine hundred thousand jobs during the ten-year time span following the signing of the North American Free Trade Agreement (NAFTA), 1994 to 2005. One-half of all of the textile jobs in the United States were once located in eight southeastern states, and in 2002 one-third of all textile jobs were located in non-metropolitan counties, which suffer some of the highest rates of poverty. Textile manufacturing in the United States has had the most geographically concentrated sector of employers in rural America.

Gathering skirted wool.

The political and cultural consequences of losing jobs that employ people in rural communities, particularly women and those without high school diplomas, have been cascading. The reality of how these changes impacted one community in Eastern Kentucky was brought to my attention during a conversation with Martin Richards, the executive director of the Community Farm Alliance based in Berea. "A generation ago Kentucky's rural communities had two things going for them," Richards told me: "Tobacco and the apparel industry. The sewing factories, largely employing women, provided a steady paycheck, family health insurance, and were often the single largest contributor to the local government tax base. Farming, typically 'man's work,' bought and paid for the homestead and supported a vast array of local businesses from the hardware and feed store to the local auto dealer and implement company. Together farming and the sewing factories supported strong local economies, put a generation through college, and created a sense of local community."

The new centralized global textile processing systems have had an impact on rural manufacturing jobs in our country and also on agricultural-based livelihoods as a whole; centralized global milling infrastructure has favored and been designed to support large-scale

After wool fleece is skirted (where less favorable parts of the wool are left behind, and the high-quality wool is preserved), the fiber is bagged and prepared to leave the farm for washing.

industrial systems of agriculture. You may think that the centralization of the value chain is somehow more efficient, and "bigger is better" has been the mantra for generations in Western culture. However, it is important to consider that in the last forty years global rates of human consumption have tripled, and there has been little to no improvement in material efficiency since 1990, meaning we now use more material per unit of GDP than we did at the turn of the century due to the fact that production has moved from what are known as material-efficient economies such as Japan, South Korea, the United States, and Europe to less materially efficient countries such as China, India, and other countries in Southeast Asia.[3]

When we ask how we can develop a new rural economic development strategy in the second half of the twenty-first century within the United States and beyond, we need to start with the question, "What types of agricultural-based livelihoods do we want to support?" For the purposes of meeting the core values that we see as integral for planetary well-being, Fibershed's concentration resides in incentivizing soil regeneration and biodiversity and establishing livelihood opportunities for young entrepreneurs who seek to stay in or become part of revitalized rural communities. It is also very important to enhance economic stability of multigeneration family-owned farming operations that are present in both our fibershed and others across the nation and globe.

When individual governments, nonprofits, and intergovernmental agencies consider how to support and facilitate the enhancement of these values, the conversation is primarily focused on a food agenda that relies on farm-to-table economic strategies. It is rare within economic development and policy discourse to see soil-to-skin strategies stacked into the economic modeling to support the types of agricultural operations that we all want to see thrive. Decentralized fiber and hide processing and milling could support our attempts to address the creation of equitable food systems. With so many producers focused on ingestible proteins like sheep's-milk cheese or grass-fed and -finished beef and lamb, we often neglect to consider that these proteins also exist as fibers derived from the same species. When we devalue one part of the animal, we diminish possibilities for enhancing land managers' income and in turn their stability. Producers benefit greatly when they can increase revenues through revaluing all of the raw materials that come from their herds and flocks. "In order for sheep production to be financially successful, the producer needs to consider the climate, the breed of sheep, and the land," said Jaime Irwin, who with her husband, Robert, offers large-scale contract grazing services throughout Northern California. "If the breed of sheep fits the climate, and the producer is able to understand, improve, and manage the land and animals, then the three things that sheep are good at—food, fiber, and land management—can provide enough revenue for a successful operation. If the focus moves from overall health and improvement of animals and land to focus on one production aspect, then loss of economic sustainability follows."

> Decentralized fiber and hide processing and milling could support our attempts to address the creation of equitable food systems.

What is the global relevance of sustaining profitable small- and medium-scale diverse family farming operations such as those that would produce fiber for the California Wool Mill? The most recent data aggregation on global food security put forth by the United Nations Food and Agricultural Organization (FAO) stated that humans consume 70 percent of their food from small- or medium-scale family farmers, and yet

In Orland, California, Sierra Reading, *left*, wears a vest she made from felted Cormo roving from Sue Reuser's sheep. Reading dyed the roving with onion skins. Reading was connected to the one-year wardrobe challenge through her California College of the Arts professor Sasha Duerr. Reading now organizes a social studies residency from her home community in Colusa, California.

subsidies (including those used to build value-added infrastructure and manufacturing systems that support farmers) largely support agribusiness. Industrial agriculture receives 80 percent of the subsidies and 90 percent of the research dollars, and between poor pay and the elimination of human labor within these systems, rural people are being driven into urban areas.[4] The report stated: "If these trends continue, by 2050, 75% of the entire human population will live in urban areas. We must reverse these trends by providing new possibilities and incentives to small and medium scale farmers, especially for young people in rural areas."

It is well known what happens when we leave the business of feeding and dressing ourselves to global-scale agribusiness: We lose touch with the inputs, the impacts, and the land itself. Food security is by far a significant enough reason on its own to justify the need to enhance economic conditions for small- and medium-scale family-owned farming operations. There is also a high value to be placed upon cultural and political stability, as many of the conflicts we see unfolding today are rooted in changes in the environment and economy that stimulate human migration and in turn lead to increased human pressure and conflict for dwindling natural resources. These cultural and political tensions are often attributed to populations moving out of vanquished rural communities.

It may not seem obvious at first to consider how intricately woven all of these pieces really are. But small- and medium-scale family farm and ranch prosperity, political stability, cultural continuity, and soil and land health are directly connected to localized and decentralized value chains that can support the efficient processing and distribution of farmed and ranched material into a local economy. The decentralization of fiber processing and milling in the mid-twentieth century is one important tool in the kit for anchoring healthy food and fiber systems within our regional economies and restoring our resource-based rural communities.

Pacific Northwest Fibershed community members work in the flax field at harvesttime, creating what are known as flax stooks. *Photo by John Morgan.*

CHAPTER SEVEN

Expanding the
Fibershed Model

*T*he Fibershed model, based on a regional implementation of the soil-to-soil framework, has expanded beyond its original geography in Northern California and can now be found in over forty locations around the world, including the United States, Great Britain, Canada, and Australia. We are excited to see a growing number of citizens focused wholeheartedly on the development of regional textile cultures and economies. It is incredibly promising to see fibershed projects arise in countries whose economies were historically founded upon, and continue to perpetuate, the values of colonization. Fibershed, the organization, is dedicated to nurturing and supporting these nascent efforts, helping to turn little "bonfires of change" into bigger ones, as well as sharing their stories of success. For those organizations that want a more formal relationship, we have implemented the Fibershed Affiliate Program, sharing the Fibershed logo and website template with members as a way of contributing to the growth of the regional textile movement. In this chapter I will share some of these success stories.

The Fibershed Affiliate Program

The Fibershed Affiliate Program was developed out of our efforts to develop our own regional membership-based producer directory within fifty-one counties in the Northern California fibershed. The mission of the directory is to create a network of farmers, ranchers, designers, sewers, weavers, knitters, mill owners, natural dyers, and others so they can connect with one another as well as meet potential customers. If a farmer or artisan is involved in directly regional agriculture or value-added processing, keen on upholding the values of a local and fair economy and ecological balance—then they can become a certified member of the directory and use our logo at their place of business or in marketing materials if they choose. The directory has proven popular, growing to more than 170 members in just a few years. The Fibershed Affiliate program extends this idea globally, with the goal of creating an evolving network of community-based regional fiber systems centered on information sharing, connectivity, and carbon farming practices. Thanks to a new web template tool and a series of online gatherings, affiliates are starting their own producer directories while promoting educational events and sharing fiber systems stories and research on project pages and blog posts.

A frequent question we get is, "How do I start a fibershed where I live?" Not surprisingly, we learned that regional fibersheds share common challenges, including the loss of textile infrastructure, a dearth of mills, underdeveloped supply chains, and a lack of educational opportunities for community members to learn about carbon farming and the soil-to-soil model. To help overcome challenges we created a series of webinars to share research, information, and lessons learned. We also implemented a micro-grant program ($2,500 to $3,000) to provide seed funding for new projects at affiliate members. Although groups can use the Fibershed logo and website template if they are in alignment with our goals, there isn't a formal certification process. We understand that customized versions of the Fibershed model are required to fit local conditions. "We're building and strengthening decentralized systems," said Jess Daniels, the communications director for Fibershed. "We don't want to tell New England or anywhere else how to do a Fibershed, but we can share experiences." Not all groups succeed. Some flare out, some merge with other organizations, and some don't see as much growth as they would like. But we are witnessing many affiliate members blossom, which gives all of us a great deal of inspiration and joy.

Commercial Flax Returns to Oregon

Pacific Northwest Fibershed was founded in 2013 by Shannon Welsh, a textile designer and educator based in Portland, who originally hails from her family's flax farm in North Dakota. The organization's goal is to foster the re-emergence of a bioregional textile community that will support soil-to-soil processes in fiber, textile, and garment production. Its geographic limit is a three-hundred-mile radius around Portland from which all natural fibers, dyes, and the labor needed to create the textiles

Shannon Welsh and Angela Wartes-Kahl with their 2017 flax crop at harvesttime. *Photo by Micah Fischer.*

are to be sourced. Initially, the organization's focus will be on linen. Too cold to grow cotton, western Oregon's climate is very amenable to flax, which can be grown in the summer without the need for irrigation. The Pacific Northwest Fibershed's linen project will focus on the dual-purpose flax crop (seed and fiber) known as Linore, as well as fiber cultivars such as Ione and Agatha.

Oregon was once a leader of flax production in the region, particularly the Willamette Valley, which has soil and climate conditions similar to the centers of European flax farming in Flanders and the Low Countries. In the years prior to World War Two, thirty-seven thousand short tons of flax were grown annually in Oregon on nearly twenty thousand acres of farmland, enough to keep fourteen mills in operation. There was an annual Flax Festival in Mount Angel, located halfway between Salem and Portland. During the 1940s, demand for flax fiber jumped to meet a rising need for clothing, tents, and rope for the war effort. Then production crashed as Europe began growing flax again. Meanwhile synthetic fibers, such as nylon, began to make their way into American clothing, further reducing demand for linen. By 1960 all fourteen mills had closed, and Oregon farmers quit growing the plant for commercial fiber use. By 1980 nearly all of the region's textile manufacturing infrastructure had deteriorated to the point of uselessness or had been sold off. By 2015 the only flax plants being grown were for seed production or for research purposes.

For Shannon Welsh and Angela Wartes-Kahl, an organic farmer in Alsea, Oregon, the long history of flax production in the state meant the time was ripe for a resurgence of the bast fiber. For Wartes-Kahl, the initial motivation for her involvement was straightforward: She wanted a locally sourced linen work shirt to wear during the hot summer months. Together the two women founded Fibrevolution, a business dedicated to bringing back flax. Working with Oregon State University, a natural fiber consulting firm, local farmers, and a curator at the Willamette Heritage Center, Fibrevolution coordinated the successful planting and harvesting of four acres of flax—harvested by hand—signaling the early stages of flax's return to Oregon. In 2018 the farming effort expanded to eighteen acres. A research trip to Europe by Welsh and Wartes-Kahl to meet with seed companies, search for effective processing equipment, and visit a flax farm cooperative was also successful. A grant from Patagonia's Environmental and Social Initiatives Program to Pacific Northwest Fibershed has also helped the flax project get off the ground. They hope all this work and research will eventually lead to the construction of a modern flax processing facility in the area.

They discovered one major mechanical challenge: finding harvesting equipment that can pull the plant, not cut it, which ensures that the maximum amount of high-quality fiber is available for its eventual transformation into linen. Unfortunately none exists in the United States—none that works properly anyway—which means that so far, plants must be harvested by hand on Fibrevolution's farm plots. Despite the many challenges, Welsh and Wartes-Kahl remain undaunted in their effort to bring back flax as a viable commercial crop in Oregon. They view linen as a perfect fit for the Pacific Northwest, complementing the abundance of wool and other animal fibers suitable for cool weather and already available in Oregon. "We see flax as a very promising crop for re-introduction into our agricultural system," Welsh wrote on the Fibershed website. "The crop can be grown organically and has the potential for protecting our natural environment, and reducing health risks for farm workers . . . [it] can be drilled without much need for tillage, which enhances the soil building abilities of this crop."

Alpaca Farming in New England

Alpaca farmers are coming together in Massachusetts. Founded in 2017, the Southeastern New England Fibershed is a network of fiber farmers, textile and apparel designers, manufacturers, brands, advocates, and educators that operates within a one-hundred-mile radius around the historic textile manufacturing centers of New Bedford, Massachusetts, and Providence, Rhode Island, cities with extensive textile infrastructure still in place. The organization's mission is to create a regional fibershed that unites fiber to finished product by connecting the dots of the supply chain to bring production back and reinvigorate a once thriving New England textile industry. At its height in the 1880s, textile manufacturing employed tens of thousands of workers in the region; it then entered a precipitous decline that saw nearly all the jobs leave by 1950. However, a recent resurgence in manufacturing led by boutique clothing lines, natural fiber entrepreneurs, and hipster designers has led to the creation of new factories and new hiring in the area. As proof of this resurgence, the value of textile goods made in Massachusetts rose from four hundred million dollars in 2009 to six hundred million in 2013, according to the US Department of Commerce.

In a bid to assist this burgeoning development, the Southeastern New England Fibershed requested and received a grant from our organization to begin developing a carbon farming cohort centered on alpaca producers in the area. The cohort includes six alpaca farmer members and has already connected with the New England Alpaca Fiber Pool, a company founded by three fiber artists in 1997 to aggregate small amounts of alpaca fiber from farms across the region into commercially viable quantities. Everyone who joined the cohort was interested in discussing the carbon farming poten-

> The Southeastern New England Fibershed requested and received a grant from our organization to begin developing a carbon farming cohort centered on alpaca producers in the area.

tial of their grazing and animal management systems and exploring next steps as outlined in our Climate Beneficial Transition Verification document. Working with a pasture management consultant and soil scientists at the Woods Hole Research Center, the alpaca cohort will set the stage for producer applications to the Massachusetts Department of Agricultural Resources' Agricultural Environmental Enhancement Program.

According to its website, Southeastern New England Fibershed hopes the cohort's activities will create "Carbon Farming resources that are unique to alpaca farmers and can be a model for alpaca growers elsewhere in our Fibershed and nationally." They also envision that the carbon farming plans created through this project will represent approaches that are appropriate for the Northeast climate and vegetation more generally.

The seeds of the Southeastern New England Fibershed were nourished by an all-day roundtable in 2017 hosted at Joseph Abboud Manufacturing in New Bedford, which drew seventy-five people to explore the extent of area's production capabilities, examine opportunities for collaboration, and identify barriers and next steps. Participants included farmers, fashion designers, elected officials, agriculture advocates, textile artists, and representatives of garment manufacturers. During the event, they were pleased to announce a "Farm to Fashion" collaboration with the Rhode Island School of Design in Providence. After the roundtable discussion ended, the group took a tour of the Abboud factory, one of the largest apparel production facilities in the United States, employing over eight hundred garment workers.

Since formalizing the Southeastern New England Fibershed, co-organizer Amy Dufault has been spearheading gatherings of regional Fibershed organizers. Connecting across the Greater New England region, Amy has been facilitating meet-ups and co-presentations with nearby groups including the Western Massachusetts Fibershed, Connecticut Fibershed, and efforts like Local Fiber New York and the Textile Lab organization that has been cataloging regional fiber production and has been a member of the Fibershed Affiliate Network since 2016. Working within ecological and cultural microclimates of the densely populated northeastern United States, these gatherings emphasize the power of building momentum through a mission-aligned and decentralized network.

The Bristol Cloth Project

A mill for weaving cloth, the first in ninety years, has opened in Bristol, England. The Bristol Weaving Mill was the inspiration of Juliet Bailey and Franki Brewer, natural fiber artisans and entrepreneurs. They launched the micro-mill in 2015 in the heart of the city's old textile district with the goal of creating opportunities to produce innovative, high-quality woven fabric using a traditional British manufacturing process—but with an environmental mission. The mill, which features a refurbished Dornier dobby loom purchased secondhand in Holland, is available for projects by other fiber artisans and can handle many different types of

The Bristol Cloth project has developed a regional supply system to grow, dye, and weave a woolen cloth in southwestern England thanks to the contributors Fernhill Farm, Botanical Inks, and Bristol Weaving Mill. *Photo by Landlore.*

Dyeing at the Bristol Cloth project. *Photo by Landlore.*

raw material, though it is primarily focused on woolen products, including top-shelf tweed. From the start, Bailey and Brewer wanted the facility to create environmentally and ethically responsible garments using local wool, spun together with recycled fibers whenever possible. They wanted an ethos of accountability, traceability, and sustainability to be embedded into every bolt of cloth and garment produced at their mill.

One of the principal clients of the mill is the Bristol Cloth project, within the South West England Fibershed Affiliate region. Its goal is to showcase locally produced, holistically farmed, and biologically washed wool as a way of demonstrating that it is commercially possible to make cloth while sustaining British tradition, culture, community, land, and the economy. Bristol Cloth sources its wool fiber from Fernhill Farm, located fifteen miles south of Bristol in the Mendip Hills, a holistically managed enterprise whose land management practices regenerate soil fertility while meeting the highest standards for animal welfare. The farm raises Shetland-Romney sheep for both meat and fiber, which is unusual because the wool produced by sheep meant for the meat market is often considered to be too poor in quality for clothing. (This is a

Finished fabric from the Bristol Cloth project. *Photo by Landlore.*

similar situation to the United States, where the wool is considered usable only for industrial purposes.) Most often, it is simply thrown away—which is a cost to the farmer. The Bristol Cloth project, in contrast, considers the woolen yarn produced by Fernhill to be desirable for its quality as well as its environmental ethics.

The Fernhill yarn is naturally dyed with certified organic plant dyes by Botanical Inks, a Bristol-based company owned by Babs Behan, a natural-dye artisan and a guiding force in the Bristol Cloth project. A transplant from London, Behan was impressed both by the sense of community among Bristol's fiber community and by the deep sense of textile history that still permeates the city. The mill, she noted, is located near the site of the Great Western Cotton Works, a major industrial factory that opened in 1838 and operated until 1925, employing over a thousand workers, mostly women, at the peak of its production. For Behan, the Bristol Cloth project is a reinvigoration of time-honored textile-making methods with environmental and social justice concerns added in. It is a model, she believes, that can be easily replicated outside of Bristol and around the planet.

Three Rivers Fibershed

Founded in 2015 by Maddy Bartsch and Lydia Strand, educators and fiber activists, the Three Rivers Fibershed is centered in Minneapolis, Minnesota, and extends 175 miles in each direction, taking in portions of Iowa, Wisconsin, and South Dakota. It encompasses the Twin Cities, suburbs, small towns, rural communities, and everything in between, much of which is farmland. The fibershed's challenge is connecting such disparate resources, businesses, producers, and consumers together in a way that supports a healthy, sustainable fiber economy. To do that, Strand and Bartsch initially focused on empowering individuals from within their communities to meet the unique needs of geography.

In 2017 Fibershed provided Three Rivers a micro-grant to support a self-designed project that aimed to help farmers and ranchers who felt isolated gain access to needed resources and make contact with fellow fiber producers to build a supportive network. That means helping producers find access to markets in cities and other places that can bring new revenue to a farm or ranch. The grant supported a number of workshops focused on articulating the fibershed's collective strengths and challenges, including marketing, branding, product presentation, telling the story of the farm and its fiber, discussing best grazing practices, finding a shearer, livestock parasite management, which fiber animals are best for climate, and establishing a relationship with a mentor. The needs are great, Strand and Bartsch discovered. The lack of direct-market opportunities for local fiber is significant, as are the challenges connected to the lack of effective branding and packaging of fiber to pique consumer interest, especially since many potential customers may be reluctant or unwilling to pay a premium price for local fiber. Currently the Three Rivers Fibershed is compiling a producer directory, similar to what we have done in Northern California, in order to better connect producers and consumers.

Recently Three Rivers conducted a workshop on a "One Year One Outfit" challenge (similar to my own one-year wardrobe challenge) where individuals commit to making a locally sourced outfit in one year using the Fibershed principles of Local Fiber, Local Labor, and Local Dyes. The goal of the workshop, the first of four, is to facilitate the formation of a group whose members support one another in working to create an ethical and sustainable outfit on a budget. The idea for the challenge originated with Nicki Colls, a sewing expert and textile artisan in Melbourne, Australia, who became curious about fabric supply chains and decided to spend a year sourcing one local outfit. The resulting project,

Nicki Colls wearing the garments she created for the original One Year One Outfit challenge in Australia. *Photo by Kerry Bardot.*

which quickly became part of our Fibershed Affiliate Program, has invited people from around the world to participate, including Bristol Cloth, cultivating a diverse range of local clothing and community. The rules of Nicki Colls's One Year One Outfit challenge are simple. An attempt to create an outfit should fall into one of three categories: master, reinventer, or investigator.

MASTER

- The fiber must be farmed and processed wholly within a set radius, though some remote manufacturing is allowed where it is not available locally.
- All fibers must be natural.
- Any dyeing must use local non-synthetic materials.
- All fabric and clothing made must be of quality construction so as to ensure the life of the clothing is long and not need excessive ironing or washing.

REINVENTER

- The main fiber must be from your fibershed.
- The outfit can include secondhand components or notions (zippers, interfacing, lining) where local options are not available.

INVESTIGATOR

- Any fiber or notion is permitted, but you must investigate the entire supply chain and tell the story of why you chose it.
- You can use secondhand components, but again, tell the story of where it was made and what it was made from.

Colls's challenge resonated with Maddy Bartsch of Three Rivers Fibershed, who three years ago gave up buying new clothes that were not ethically made and sustainably sourced. "I really like personal adornment as a way to wear my values," Bartsch told a reporter. "When someone compliments me, I like to be able to share my values with them and have that clothing be a point of connection."

Upper Canada Fibreshed

The Upper Canada Fibreshed was founded in 2015 by Becky Porlier and Jennifer Osborn, farmers and fiber artists, with the goal of nourishing an emerging bioregional textile community of producers and consumers that

This queen-sized custom three-panel blanket weighs eight pounds. It was made for a patron of the Upper Canada Fibreshed circa 2015: Handwoven by Deborah Livingston-Lowe, Upper Canada Weaving, with a blend of Shetland from Chassgne, Gotland from Robin Theon, and alpaca from Freelton Fibre Mill, spun at Freelton Fibre Mill and Wellington Fibres. *Photo by Becky Porlier, modeled by Mike and Jay Pieterse.*

value sustainable agriculture and hyper-local textile manufacturing within a 250-mile radius of Toronto, Canada. Like other affiliates, the Upper Canada Fibreshed supports a grassroots network that promotes the development of a regional fiber system through education, research, peer-to-peer knowledge exchange, and building an inventory of materials, skills, processing equipment, and expertise. Their educational message is on target: Regionally produced, all-natural textiles can reduce CO_2 emissions, eliminate toxic by-products of synthetic fabric production, discourage water pollution, and encourage a culture of ethical consumption. They believe that scalable, innovative, and regionally specific value chains have important and positive impacts on producer well-being and prosperity. "Regenerative fibre farming systems, within local processing value-chains, have the potential to create economic opportunities for rural communities by meeting the growing demand for local, sustainable textile products," said Porlier. "Investing in these systems returns carbon to the soil [and] protects our water and foodsheds, all the while encouraging meaningful work in a highly skilled and collaborative field."

Becky Porlier and Jennifer Osborn met at an organic conference in Guelph, Ontario, and they both had a strong interest in the Fibershed model. After drinking too much coffee, by their own admission, they decided to jump into the current Fibershed model with both feet—as long as they were allowed to use the Canadian spelling of *fibre* (we were happy to agree). One reason they chose such a wide radius around Toronto was the relatively few fiber producers in the province. There are even fewer processors and manufacturers in this region, so to bring as many resources together as possible their Fibreshed was made quite large. Since then additional affiliate members have come into existence in Canada, including the Sunshine Coast BC Fibreshed, located on a hundred-mile stretch of British Columbia's western coast near Vancouver; the Vancouver Island Fibreshed, focused specifically on the hyper-local movement in their community; and the Pembina Fibreshed in southern Manitoba.

One of the focuses of Upper Canada Fibreshed has been on household wares such as blankets and other durable textiles that would be better suited to the coarse wool available in Ontario. Osborn explained that "textiles are so ubiquitous in our daily lives, we barely notice them. Curtains, bedsheets, carpets, upholstery fabric; it is everywhere. Natural fibres are also used in oil spill cleanup, sound insulation, bedding, stuffing, canvas, and so much more." Some of these things could be manufactured within Ontario with the wool that is already available.

> One of the focuses of Upper Canada Fibreshed has been on household wares such as blankets and other durable textiles that would be better suited to the coarse wool available in Ontario.

Jennifer Osborn has a mixed flock including Shetland sheep at her farm All Sorts Acre, located outside of Orangeville, Ontario, which she cross-breeds with Romney, Icelandic, and Bluefaced Leicester for both meat and wool. She is also endeavoring to be the first ecologically based commercial natural dye farm in Ontario. "Much research went into plant selection for the gardens. Things such as plant yields to space needed, ease of growing, years needed before harvest, is the plant perennial or annual, cold tolerance, facilities needed to grow plants (we didn't want to heat a greenhouse), what mordant was needed to get the desired colour, lightfastness, potential invasiveness, plant preparation to get the colour, and of course the potential colour yielded all needed to be considered."

Upper Canada Mercantile, a member of the Upper Canada Fibreshed, set out to create a blanket that would represent the land, climatic, and human-based interactions that make a product unique, similar to the now familiar concept of *terroir* used in conjunction with French wine. Like all agricultural products, this blanket has a terroir unique to its region and people. The fabrication of these soft wool blankets depends on a wide range of talent and resources within the Fibreshed, from shearing to processing and weaving—an excellent example of what fibersheds are all about.

The Upper Canada Fibreshed also has a fashion component and has partnered with designer Peggy Sue Deaven-Smiltnieks to create garments that reflect the uniqueness of the region. Raised in California, Peggy Sue graduated from the Rhode Island School of Design with an MFA in apparel design and started her career in New York in supply-chain management and product development, designing lines for various big-box retailers. There she learned about industry-standard global supply chains, which focus on maximizing efficiencies, cutting costs, and expediting time lines. While attending seasonal textile trade shows, she noticed there weren't many North American companies. The reason? Smaller North American fabrication businesses were not developed enough to go to these trade shows and connect with established design houses.

"It was apparent that even if a consumer wanted to buy quality pieces that were traceable and sustainable, there were incredibly limited options," said Deaven-Smiltnieks, now based in Milton, Ontario. "A lot of them weren't from North American supply chains." So in 2014, post-recession, when she started conceiving her own line, she handpicked North American producers and built her own supply chain. Today her Canadian-made Peggy Sue Collection is composed entirely of North American–produced fibers. Skirts, coats, and jackets in upcycled woven denim, organic cotton, or 100 percent local alpaca and wool overturn the prevailing idea that sustainable fashion isn't sexy. All the denim used has been diverted from Ontario landfills, and there is zero waste. Even denim leftovers from other garments are blended into white wool yarn for sweaters.

Designers play a strong role in shifting practices, she contended, as do the consumers in demanding better practices. Peggy Sue said she uses North American fiber because she believes "in the longevity of the North American farmer and preserving the genetics of a varied fiber landscape. If designers do not use these fiber types for everyday clothing, there will be no demand for these farmed fibers, and [it will not be] financially possible for farmers to continue growing/raising such a variety." Or as she likes to say, "No Farms, No Fashion."

New York Textile Lab

The nine million residents of New York City depend almost entirely on an extensive and complicated delivery system for their water, a system that originates among the farms, fields, and mountains of upstate New York. Years ago city administrators faced a vexing crisis: how to ensure high-quality drinking water from remote surface sources that were essentially unregulated. One proposed solution was to build a very expensive filtration system that cleaned the water before it entered the city. A second solution involved dealing with pollution at its various sources, either through legal and regulatory confrontation with landowners or by cooperation. New York authorities choose cooperation. The result was a landmark agreement in 1997 that committed state and federal agencies, landowners, environmental groups, municipalities, and others to protecting the city's watershed. This included supporting the agricultural community, which is where the Watershed Agricultural Council came in. It has worked to reduce pollution while supporting family farms in the region through a variety of programs, including the new carbon farm planning and implementation process, that is being pioneered with fiber farms that are collaborating with the New York Textile Lab to develop regional Climate Beneficial fiber sources.

Founded in 2015, the New York Textile Lab brings together designers, family farms, and mills in New York to help grow an economically diverse, statewide textile supply chain. Like other Fibershed Affiliate organizers, designer and activist Laura Sansone recognized the significant flaws in the conventional textile manufacturing industry, including its contributions to global warming. Her response was to help create distributed networks of small businesses that include fiber farms, mills, and manufacturers with the goal of building a Climate Beneficial textile economy in the state. "When I was a student we learned about the innate characteristics of fiber materials," she wrote on the lab's website. "But we didn't really learn too much about exactly where the materials came from, particularly industrial textile materials and clothing. There was a glossed over mention of the flax plant, the sheep, and the silk worm, and there was little attention dedicated to the actual farm or the geographic region where these materials grow and the farming communities that cultivate them."

To remedy this situation, she believes it is imperative that designers identify the provenance of the materials they use as well as the social and ecological impacts of producing a textile. A particular concern is the lack of diversity in material choices, the consequence of monopolization of

At Textile Lab they use harvested plants from local markets to practice ecological methods of creating, dyeing, recycling, and composting textiles. *Photo by Laura Sansone / NY Textile Lab.*

Textile Lab was born in 2009 out of the GrowNYC Greenmarket system. It uses a mobile kitchen and workstation to research local fiber systems and environmentally ethical textile production. *Photo by Laura Sansone / NY Textile Lab.*

the clothing industry by cotton and polyester, and the downsizing that has taken place in the family farming community as a result. The Fibershed model immediately appealed to her with its emphasis on biodegradable materials, better farming practices with regenerative outcomes, reduced consumption of resources, and a focus on local jobs. That's why the lab's New York Native yarns, textiles, and products are sourced, spun, knit, and woven within three hundred miles of New York City. They are a blend of Corriedale wool from the Southern Adirondack Wool Pool and alpaca from Little Creek Farm. Another effort is the lab's Regional Cloth Project, which features textiles made from blended, machine-grade yarns sourced from the southern Adirondacks along with premium alpaca.

Strategies for Sustaining Regional Fiber Systems

From the experience gained by national Fibershed organizers, we believe the strategies required to develop and sustain regional fiber-and-dye systems are supported greatly by focusing on the following.

#1: Investing in People

There is a real need for training in areas such as classical plant breeding; fiber grading and handling; and milling processes including fiber cleaning, combing, washing, spinning, natural dyeing, knitting, weaving, and ecological finishing of textiles. Garment creation processes will require skill building in patterning, cutting, and sewing. Those hoping to develop new vocational textile skills will also need educational services and training to develop financial literacy, business planning, and administrative experience. Small business development skills would initially include a foundational understanding of business architecture, such as exposure to and awareness of shared ownership models that include milling and agriculture producer cooperative business design. Legal services offered to those who graduate from business development training would provide skilled entrepreneurs the support required to launch their projects into fully formed legal entities. These educational services could be provided by public re-investment in our junior college system.

#2: Investing in Economic Analysis

Developing infrastructure feasibility studies is a first step prior to implementing on-the-ground infrastructure. Providing communities the time necessary to map their fiber supply and existing fiber processing talent pools is the foundation for knowing where to put a mill and who might want to work in it. This is an investment that philanthropic foundations are well suited to support.

#3: Investing in Infrastructure and Market Development

These are the largest capital investments that a community will make, and so it is important to have an understanding of the "who, what, where, and why" well mapped prior to putting buildings, machines, and distribution infrastructure in place. The art of infrastructure planning is determining the scale of a project; and that scale is determined by supply and demand. When understanding demand and supply, it has been shown through our work that surveys and marketing research go only so far. Human relationships are the fundamental key to building stable supply and demand over the long term. It can take two to three years before a textile brand, designer, artisan, or distributer is able to develop the business architecture to create, market, and sell a "regionally grown and sewn" good. Hundreds of hours of calls, emails, and supply-chain support are often required to support companies to reshore. As this infrastructure re-investment is embarked upon, our communities will require large-scale financial investment, including the following:

MUNICIPAL BONDS: There are currently municipal bonds written and voted upon for investments that span the gamut from water treatment systems to mountain biking trails. Climate Beneficial clothing is arguably as essential as mountain biking trails and possibly close to on par with water sanitation processes.

CURRENT PRIVATE-SECTOR SOCIAL IMPACT INVESTMENTS: We've seen large-scale investments in technologies that have not yet been mapped for their soil-to-soil impacts (synthetic biofibers and dyes, electronic sensor clothing). These same investment sums could be rerouted to build out natural fiber and dye processing systems that have known, predictable, and safe soil-to-soil impacts.

COMMUNITY DEVELOPMENT FINANCIAL INSTITUTIONS: CDFIs provide capital to generate economic growth and opportunity in distressed and low-income communities. These federal funds are issued through mission-driven certified financial institutions and are designed to stimulate and create jobs, build homes, and create infrastructure. CDFIs are potential lenders to support the redevelopment of necessary milling infrastructure and could step in to invest after the point at which a community has had time to articulate the gaps between supply and demand for regional farmed fiber goods and textile.

USDA PROGRAMS: These include the Value Added Producer Program Grant and the Working Capital Grant. Both grant opportunities are issued through the USDA's Rural Development offices. They support our fiber and food systems. They could receive much greater funding to serve the appetites and needs of rural communities.

State and regional investments in climate change resiliency strategies and climate change amelioration are currently being generated through cap-and-trade bills and other forms of taxation and re-appropriation of funds. In 2015 our California cap-and-trade bill distributed $2.2 billion for investment in electric cars, high-speed rail, municipalities, cities, water districts, and other similar agencies and organizations.

A small portion of these funds have been allocated for our state's new Healthy Soils Program. If we are going to support the advancement of our climate work, we must expand our investment focus to holistically designed soil-to-soil food and fiber projects. Directing regional and state funds to develop the infrastructure for these systems, including regional mills and processing centers that keep raw material within rural communities until value has been added, will help to establish prosperity in these communities while reducing the ecological footprint of consumption being generated by population centers that currently rely on end-products that are largely being imported from overseas but could come from neighboring rural communities.

With appropriate investment from the public and private sectors, we can foster and grow a new set of nourishing traditions that restore functioning and healthy fiber, food, and natural dye systems. These "shed" economies can be enhanced to create permanent and lasting systems of production that sustain meaningful livelihoods within our communities. As we specifically reinvigorate fibersheds and take full responsibility for a garment or textile's life cycle, we will naturally diminish the ecologic and human rights pressures experienced by regions of the world in which production has landed for the sake of making money for a few at the expense of many.

Deepa and Loka Natarajan investigate a coreopsis flower in their dye and food garden in Berkeley, California. Deepa is an ethnobotanist, natural dyer, and the program coordinator at the UC Berkeley Botanical Garden. *Photo by Alycia Lang.*

A Future Based in Truth

Rediscovering and forming a relationship with the landscape that defines a textile resource base is a process that sensitizes us to the conditions our lands, communities, and local economies are currently in. And when we tune into these conditions, the imperative to restore ecological function to our regional landscapes and repair our social fabric becomes evident. Addressing environmental and social restoration through the lens of fiber and dye systems provides direct implementation opportunities that have been noted throughout the book. We've shown that anyone who is keen on getting involved has many early stage ways to do so—starting a dye garden in a few clay pots, using locally grown yarn from climate-benefitting farms and ranches, and mending your existing clothing. However, when it comes to launching the larger milling, textile distribution, and capitally intensive initiatives required to fully decentralize a fiber and dye system, change moves more slowly.

While we understand the benefits to decentralization and local economic empowerment, there are no immediate economic incentives for traditional investment capital to fund projects that don't include an obvious exit strategy. You can't simply exit a place-based textile culture, and therefore the economic effort to expand and enhance the functionality of our natural fiber and dye traditions hasn't been the easiest of roads within the current economic system.

The production processes that clothe the vast majority of humans today are backed by global investment capital that has been working for centuries to centralize and privatize the genetics and the machinery that undergirds fiber-dye farming and manufacturing. This system is clearly at an inflection point as we're reaching a threshold for how long centralization can exist on a planet experiencing climate disruption. Current textile systems are pumping out and overproducing materials at the expense of Earth's biomes, and it's now an economic privilege to have access to nontoxic clothing and clean food. So the obvious question arises: *Why are we still allowing this kind of economy to exist when it has made making healthy textile (and other) choices too expensive for the vast majority of us?*

To transform any kind of dysfunctional relationship and ensure solutions are enacted and mistakes aren't repeated, it's essential to understand that relationship's history. As I write this, I'm in an office built on land that was parceled out by Europeans who took the North American continent by force, carrying out genocide on indigenous civilizations that held and continue to carry forward advanced fiber and dye traditions. So advanced are their methods that textile artisans in my region are known to enhance biological diversity in a zero-waste process with tending and harvesting regimes that yield raw materials for clothing and other necessities. And in the southeastern United States, land was cleared for agriculture, and chattel slavery was imposed, a system based on stealing farmers, engineers, midwives, herbalists, and builders from many African societies. These individuals and families had pioneered indigo, cotton, and rice agriculture systems in their homelands. The enslaved agrarians were so productive and skilled that by 1836 more than half of all economic activity in the US ($600 million), was generated by the cotton industry that was operated by over one million slaves who received no economic return for their labor.[1] The income generated by slave-led cotton systems paved the way for America's rise to economic dominance.

While this story is not new, the way it has been taught to many of us is that these crimes against humanity are somehow behind us and that our economy once based in violence and extraction has somehow transformed itself over the decades and years. But if you've ever tried to garner investment from a bank or a traditional lender to rebuild or grow a new fiber- and dye-based business that is focused on improving the social and environmental conditions of your very own community, you run into the same mechanisms of economic dysfunction that profited off of genocide and chattel slavery. The US economic system depends on quarterly measured rates of return that more often than not exceed the rates of Earth's ability to produce raw materials. These rates of return are known to drive exploitation of human, plant, and animal bodies—our economy's "success" is forcing us into ecological and social debt. At this moment trillions of our own dollars are holding up this failing system; retirement funds (and most if not all of our fiduciary capital), private equity firms, hedge funds, and most of the private banks where we store our money rely on rates of return extracted from projects and endeavors that are creating long-term impoverishment for us all. A sane approach to economic development would include building a new system based on incentivizing us to focus on the regenerative capacity of our land and physical bodies.

Despite the path that the global textile economy has taken, our community is activated to change this system; allies from all walks of social and economic life have begun to work together to see this through. Our fibershed continues to carve out opportunities and identify the constraints of what a regional fiber and dye system can produce. We've become a veritable think tank of social, economic, and policy data over the last ten years. We have also engaged in the fundamental work of generating raw materials from integrated and regenerating fiber, food, dye, medicine, and fuel systems. We have just begun to scratch the surface of all the potential milling systems that are available to turn what we grow into beautiful textiles.

We have also been able to elevate the voices and faces of the farming, ranching, and artisan community by bringing their stories into the public eye. We no longer experience fiber as simply a commodity, nor are dyes an anonymous chemical cocktail. We've spent years connecting these once unknown materials and processes to actual lived human experience, and in doing so we've begun to positively revalue the role and economics of fiber and dye farming. As my colleague Dan DiSanto says, "The future is in reversing the supply chain." This means putting our attention on the fact that textiles come from the soil and that there are people to acknowledge who wake up each day committed to their land, plants, and animals. People who enjoy being connected with others who appreciate and utilize the fruits of their labor within their own work and life practices.

In the textile farming and making community in my region, the stories have largely been focused on the experiences of women like myself—those whose families immigrated over the centuries to colonized lands, bringing their farming and design practices with them. The movement to fully decentralize and regenerate our textile systems includes holding up and placing long overdue value on and respect for indigenous and folk traditions—especially those held by people of color whose stories have and continue to be erased by fast fashion and industrial agriculture.

The next phase of this work is to establish that the local fiber and natural dye work can expand to serve and benefit more than the privileged few. To ensure that this occurs, many aspects of the system need to be transformed, one of the most significant being how investments are made in milling and distribution systems. The future of these textile systems could become climate-benefitting and economically empowering for all those who work along the soil-to-skin pathway. This emerging future is already starting to show itself, and all of us have a role in hastening its birth through our voices, actions, and consistent care.

Craig Wilkinson plants indigo with a volunteer team. Craig provides seeds and dye workshops for the Northern California Fibershed community. *Photo by Kalie Cassel-Feiss.*

Acknowledgments

Thank you to all the farmers, artisans, small mill operators, shearers, ranchers, and makers in the Northern California Fibershed who inspired so much of this work, and whom I witness day in and day out working to revitalize and craft their lives around and from their innate human creativity and passion to live close to the Earth and be of their place. A special acknowledgment and thank you to sheep flock owner and farmer Jean Gowan Near who passed at the ripe age of 105 during the process of writing this book. She taught me so much about the way that things were done not so long ago, and how true sustainability looked, felt, and functioned. —RB

My work on this book would not have been possible without the financial support of the following: the Bradshaw-Knight Foundation, the Bybee Foundation, the Grasslans Foundation, the Globetrotter Fund, the O'Toole Family Foundation, the Regenerative Agriculture Foundation, and the Stokes Foundation. I deeply appreciate the support of every individual involved with these organizations. —CW

The Irma Jumpsuit, designed by Geana Sieburger of GDS Cloth Goods for Fibershed's Climate Beneficial Gala, was made with Community Supported Cloth of Rambouillet wool from Lani's Lana, woven by Huston Textile Company. *Modeled by Serah Blackstone-Fredericks.*

Ingredients to Watch

There are numerous and shifting numbers of synthetic compounds utilized in textile manufacturing. The American Association of Textile Chemists and Colorists Buyer's Guide lists two thousand chemical specialties in over one hundred categories—and this does not include textile dyes. In an effort to isolate those that are known to have the most severe and well-documented health impacts, a subset of compounds has been compiled here. While these compounds and their effects are listed singularly and linearly, when inhaled, ingested, or absorbed these chemicals act synergistically; it has been shown that the combination of compounds has a greater detrimental impact than any one compound working alone.[1] It is critical to understand that research continues to uncover that "allowable levels" of these compounds often exceed what many of our bodies can withstand, causing disease or chronic illness at much lower thresholds than what is permitted within the industry's standards.[2]

During third-party laboratory analysis of products coming from major brands, clothing and accessory products in some cases were found to have compounds of concern at concentrations beyond the company's stated allowable levels.[3] This can occur due to the sprawling and shrouded nature of the global supply chains that the industry relies upon to produce its branded products. From research done at the time of publication, the one global standard that provided protection via the elimination of the synthetic compounds mentioned below is the Global Organic Textile Standard (GOTS). This standard goes into depth in regard to a myriad of compounds of concern, far beyond the short list presented here. The standard does allow for the use of synthetic dyes, but it has eliminated the use of azo dyes (which are highlighted below).

Phthalates

Diesters of phthalic acid are commonly referred to as a class of chemicals known as phthalates and are frequently used to make plastics softer, among other applications including plastic inks for textile screen printing. Plastisol inks containing phthalates are the most ubiquitously used material for textile printing and were found in thirty-three of the thirty-five children's clothing items tested by independent third-party labs through Greenpeace's 2014 study *A Little Story About the Monsters in Your Closet*. In a 2016 study titled *A Red Card for Sportswear Brands*, phthalates were found in the soccer shoes, gloves, and jerseys sold as part of World Cup merchandising. The compounds are not chemically bound to the plastic and therefore off-gas for the duration of a garment's life.

These compounds are commonly found in indoor air and dust.[4] DEHP and DINP are commonly used members of this class of compounds and are produced to the tune of thirty and twenty-four million pounds per year, respectively. And so it also comes as no surprise that the Centers for Disease Control found that nearly 100 percent of the US population has measurable levels of one or more of these chemicals in their bodies.[5] Phthalate esters migrate from our garments and household products into our bodily fluids and have most notably been found in amniotic fluid, breast milk, and maternal urine samples taken during pregnancy.[6] It is highly concerning that phthalates so easily migrate to our bodily fluids and especially the fluids of pregnant mothers; the compounds have been found to disrupt and alter testis cell hormone production during a critical time of male in-utero development. Phthalates lower the testicular testosterone levels when the fetus is undergoing sexual differentiation, causing what researchers have named phthalate syndrome, which is known to create incomplete testicular descent, smaller testis weight and penile size, and a shortened anogenital distance or AGD.[7] These compounds have also been associated with "less male" play behavior in boys, and findings suggest that phthalates have the potential to alter androgen (male hormone) responsive brain development in humans. Phthalate exposure has also been associated with decreased male fertility specifically due to decreased semen quality.[8]

PFCs

Perfluorinated and polyfluorinated chemical compounds (PFCs) are synthetic (human-made) chemicals that have been in production for more than fifty years. Both the long- and short-chain PFCs are well known for their ability to persist and accumulate in the environment, due to the strength of the carbon-fluorine bond. PFCs have water- and oil-repellent qualities and are used for stain- and grease-proofing garments, carpets, and leather products as well as creating durable water-repellent gear like raincoats and jackets and mountaineering garments. FTOH (fluorotelomer alcohol), a common precursor chemical compound used for waterproofing, is known as a volatile PFC and was found to break down into PFOA (a long-chain PFC) once it entered the human body and oxidized.[9]

Governments in the United States and Europe have made efforts to phase out the production and use of the better-studied long-chain PFCs. In 2002, 3M Corporation phased out long-chain perfluorooctane sulfonic acid (PFOS), and perfluorobutyrate (PFBA) was last produced in the United States in 1998. It is, however, widely understood that these molecules are designed to persist under extreme conditions (their role is to keep a textile water-repellent, and thus they continue to function persistently within the environment even after being washed from our garments). An example of this biological persistence was made evident in a 2009 study that took place over a decade after the phaseout—researchers identified the very first PFBA umbilical cord blood sample in the United States, ten years after we'd discontinued production of the substance.[10] According to a 2014 report by Greenpeace, PFCs have been found in the livers of polar bears, in the excrement of penguins in Tierra del Fuego, and in snow samples on the remote Italian Alps—all locations where consumption and production of PFC-laden textiles do not exist.[11] However, these compounds are now found in the farthest reaches of our planet due to their ability to bioaccumulate and concentrate in our environment. Unsurprisingly, PFCs accumulate in human blood and breast milk, and PFBA was just one PFC found in the umbilical cords of newborns.[12]

Researchers at the Centers for Disease Control, Columbia University, and Johns Hopkins tested cord blood from babies born at Johns Hopkins Hospital between November 26, 2004, and March 16, 2005, for ten PFCs. They detected PFOS in 99 percent and PFOA in 100 percent of the samples. Eight other PFCs were detected at lesser frequency.[13] The Environmental Working Group's 2009 report *Pollution in People: Cord Blood Contaminants in Minority Newborns* states that researchers at the Johns Hopkins Bloomberg School of Public Health tested PFC levels in nearly three hundred mother-and-infant pairs and found that women with elevated blood levels of these chemicals gave birth to infants with reduced birth weight and head circumference.[14] Other human studies have linked PFC exposure to difficulty conceiving, lower sperm quality, thyroid disease, and elevated cholesterol.[15] In animal studies PFC exposure has been associated with immune suppression, as well as pancreatic, liver, and breast cancers. Exposure to some PFCs has also been associated with kidney and testicular cancers in humans.[16] PFOS and PFOA are negatively associated with concentrations of vaccine antibodies in vaccinated children.[17] While the United States has halted production of a subset of these compounds, PFCs continue to make their way into our homes and closets through the importation of garments and durable goods from countries whose manufacturing processes depend upon them and have no laws prohibiting their manufacture. In *A Little Story About the Monsters in Your Closet*, Greenpeace found that fifteen of the fifteen children's garments tested positive for one or more PFCs. In the study *Chemistry for Any Weather*, also published by Greenpeace, fourteen out of fourteen rain jackets manufactured by the world's leading outdoor brands were found to have traceable levels of PFCs. There is currently no regulation inhibiting the importation of clothing that harbors these compounds. There are, however, alternatives, and searching for PFC-free garments is now possible as a growing number of companies are working to offer safe alternatives.

Nonylphenol Ethoxylates

Nonylphenol ethoxylates or NPEs are synthetic (human-made) chemicals that are used as stabilizers and emulsifiers in plastics and are widely used in detergents and as surfactants in textile manufacturing. NPEs can enter the body by inhalation of air containing the compound, through ingestion of contaminated food or water, and through dermal contact with products.[18] NPEs degrade into NPs (nonylphenols), which are known endocrine disruptors. Acting specifically as an estrogen mimic, they accumulate and persist in our environment and are known to be highly toxic to aquatic life. NPEs are used during the washing and finishing processes of textile processing and remain on our clothing after purchase. Countries that manufacture the majority of our garments (China, Bangladesh, Vietnam, Myanmar, Cambodia, and India, to name a few) have no imposed bans on the manufacture or usage of NPEs, therefore garments are imported and purchased with these chemical compounds still present and detectable—and upon laundering, the residues from our treated textiles are re-released into our local aquatic and land-based environments via circulation through our public wastewater systems. Eventually the compounds are deposited into rivers, lakes, oceans, sediments, and soils.[19] Due to a widespread understanding that these compounds move fluidly and accumulate in our water supplies, the European Union moved ahead in 2005 and listed NPEs as "priority hazardous substances" under the EU Water Framework Directive.

The latest report from Greenpeace (May 2014), *A Red Card for Sportswear Brands*, tested thirty-three products from sportswear companies Nike, Adidas, and Puma. Each of these companies had committed to eliminate NPEs and detox their supply chains two years prior. In laboratory tests of soccer T-shirts, shoes, and gloves, NPEs were found in various percentages in a majority of the items tested.[20] In 2015, a year after the *Red Card for Brands* was issued, the European Union banned the importation of textiles containing even minute residues of the substance.[21] In the United States, at around the same time (2015), the EPA instituted a new use rule for fifteen NPE/NPs, requesting that manufacturers within the United States provide ninety days' notice to the EPA prior to the usage of the compounds—providing the EPA enough time to determine, "if warranted," an ability to take action to "prohibit the activity before it occurs."[22] Needless to say, this rule does not explicitly regulate for the elimination of the domestic use of NPEs, and we remain far from banning imported textiles that contain the compounds.

While the laws in the United States remain comparatively lenient compared with those in the European Union, our Environmental Protection Agency does recognize these chemicals as posing health risks, and the agency is working with manufacturers to support their voluntary phaseout for domestic use. In 2010 the US demand for NPE/NPs was 380 million pounds. According to research aggregated by the EPA, the compounds have been found in the surface water and sediments of the Great Lakes and the Ohio River—and were also found in bottom-dwelling fish (carp). The estrogenic effects of NPEs/NPs in fresh- and saltwater ecosystems create ecological effects at each trophic level.[23] Estrogenic compounds such as NPs have been correlated with estrogen-dependent cancers in adults and children. Such cancers have been on the rise at the rate of 0.6 percent per year since 1975, and in the research study "Association of Environmental Chemicals and Estrogen Metabolites in Children," one-third of the children studied had measurable levels of NPs in their urine.[24]

BPA

Bisphenol A (BPA) is a synthetic compound most widely known for its use in the production of plastics, although it's also used in textile production. The compound was originally created at the end of the nineteenth century; today the United States manufactures 2.3 million pounds of BPA each year, and approximately 7 billion pounds are produced worldwide. The compound generates six billion dollars in earnings and is produced by Bayer, Dow, Hexion Specialty Chemicals, SABIC Innovative Plastics, and Sunaco.[25] Bisphenol A is a known estrogen mimic; the National Institute for Environmental Health Sciences has concluded that there is some concern that the compound induces abnormal reproductive system development, diminishes intellectual capacity, and causes behavioral problems.[26] Other studies have shown that BPA can induce reproductive system cancer, obesity, diabetes, early puberty, resistance to chemotherapy, asthma, and cardiovascular system problems.[27]

A recent epidemiological study by Kaiser Permanente, published in November 2009 in the journal *Human Reproduction*, showed that exposure to BPA in the workplace could have an adverse effect on male sexual function.[28] BPA has been measured at health-impacting concentrations in rivers and estuaries as well as in sediment. It has now been detected in the cord blood of nine out of ten newborns in tests performed in 2009 by the Division of Biological Sciences at the University of Missouri–Columbia. As ubiquitous as it is in our blood, BPA is also concentrating in our oceans.[29] The Centers for Disease Control has stated that the substance is now detectable in 93 percent of Americans over the age of six. Well-known biological scientist Frederick vom Saal was quoted in an interview with *Mother Jones* magazine: "A poison kills you. A chemical like BPA reprograms your cells and ends up causing a disease in your grandchild that kills them."

Synthetic Dyes

Approximately ten thousand different synthetic dyes are used on our garments.[30] Derived from fossil carbon sources, these compounds are based upon the scientific innovation of William Perkins, who discovered the first synthetic dye during an attempt he made in 1856 to create a cure for malaria. Synthetic dyes are a relic of the Industrial Revolution. The original recipe required four hundred pounds of coal tar for a four-ounce yield of blue dye. Today's synthetic dyes are responsible for the color of the world's textiles (minus a small percentage of naturally dyed and undyed garments). The most ubiquitously used family of colorants, responsible for 60 to 70 percent of the total industrial practices, are azo dyes—shown to be both carcinogenic and mutagenic (they disrupt the

DNA in a cell).[31] There are currently twenty-two azo dyes that the European Union does not allow for use on textiles that will be worn directly next to the skin. Aniline yellow was the first azo dye to be brought to the marketplace, in 1861.

More than half a trillion gallons of water is used in the textile dyeing process on an annual basis, and approximately two hundred thousand tons of dye is left unbonded to the textiles and lost to effluent.[32] When allowed to enter freshwater aquatic systems, these coloring agents cause oxygen deficiencies and can heavily impact drinking and irrigation water.[33] In a 2010 study three synthetic dyes were identified in river water in various concentrations; the resulting drinking water sourced from the river contained these dyes after water treatment had occurred, showing that methods such as flocculation, coagulation, and pre-chlorination were not enough to clean the water for human consumption.[34] Dyes do not break down easily and resist biodegradation, and for this reason they persist in our waterways.

In an interview with Dr. Shanna Swan, professor of environmental medicine and public health, obstetrics, gynecology, and reproductive science for the Icahn School of Medicine at Mount Sinai, the subject of aniline dyes came up. "To put it simply, we have far more data on compounds that are ingested—such as pharmaceuticals—because we take more responsibility for testing their impact on human health. We have far less data on substances that are not designed to be ingested, but what we have realized via extensive research is the impact of painkiller acetaminophen/paracetamol (branded as Tylenol and other name brands) on the human endocrine system," said Swan. Acetaminophen reacts with our body in the same way a phthalate does—these painkillers are known endocrine disruptors.[35] Swan went on to say, "When our body is exposed to *aniline* (a component of textile dye), it metabolizes into acetaminophen (N-acetyl-4-aminophenol or NA4AP) and is known to pass through the placenta and has been detected in human breast milk, and it is a known endocrine disruptor."

In a German study aniline was present in over 90 percent of the population.[36] What we still do not understand is the rate and impact of aniline being absorbed by the skin or inhaled via lint—these are areas for more research. The question remains: How are we being exposed to aniline? There is no oral ingestion, but urine samples continue to show evidence of exposure. Aniline's metabolite, NA4AP, has been shown to cause health problems at the most sensitive moments of human life. When a fetus is exposed to anti-androgenic endocrine disrupters, we see issues such as reduced anogenital distance, smaller or disfigured penises, and behavior shifts that show up as "less male" play behavior in young children.[37] "Eight weeks is a critical time for a fetus to differentiate itself as a male—testosterone is required for this transformation, and if there are any impediments or changes in the levels of this hormone due to a mother's exposure to these anti-androgenic compounds—it can create the kinds of physical changes and mutations we are seeing in far greater number now," said Dr. Swan.

Glyphosate

In a study published in the *Journal of Environmental Sciences*, it was noted that glyphosate, the active ingredient in Roundup, is the world's most widely used herbicide, showing up in our public parks, in our neighbors' yards, in the sidewalk cracks where little weeds once grew—and on our clothing.[38] According to the USDA, 94 percent of the cotton planted in the United States in 2015 was genetically engineered to be resistant to the use of key herbicides that contain glyphosate as their main ingredient. These commercial herbicides, in addition to Roundup, are PROMOX and WeatherMAX, and in fiber production are used to destroy non-cotton plant species in the field while leaving the farmed species present. Cotton crops account for a considerable amount of America's farmed soils; it is grown in seventeen states and covers twelve million acres of farmland. Three-fourths of this cotton is used for our garments, and 6.5 billion pounds of cottonseed enters the food system annually. We export seven billion dollars' worth of the crop, and US cotton accounts for 30 percent of the total world export market. According to the National Cotton Council, farmers spend $780 million per year on chemical inputs and $170 million on farm labor. (Chemicals now cost more than human labor.) In a study from the scientific journal *Toxicology*, it was shown that glyphosate disrupted human cells within twenty-four hours of exposure in sub-agricultural doses.[39] Due to these endocrine-disrupting properties, glyphosate increases breast cancer cell proliferation in the parts-per-trillion range.[40]

We also now know that endocrine disruption leads to metabolic syndrome, obesity, and type 2 diabetes.[41] Dr. Michelle Perro, a veteran twenty-year pediatrician and educator in our local community and author of *What's Making Our Children Sick?*, was interviewed recently on the impacts of glyphosate on her patients and said: "Glyphosate causes liver damage in the 1 part per billion range, and this has caused liver disease in one in four children, equating to a child's inability to detoxify their body properly, leading to many difficulties including the most rampant which is severe allergies." There are currently no laws regulating the concentration of glyphosate residues on our clothing, bedding, or other household textiles.

Dimethylformamide (DMF)

DMF is a compound utilized in the creation of human-made acrylic fibers. Women who work in textile factories that produce acrylic fibers have seven times the risk of developing breast cancer as women outside of the textile worker population. The Centers for Disease Control has identified DMF as present in textile dyes and pigments and warns against skin contact with the substance, stating that it can cause liver damage and other adverse health effects, including skin problems, and can alter the genetic makeup of a cell, meaning the compound is known to be mutagenic. The CDC has also identified that over one hundred thousand workers in the United States are exposed to the toxin—a colorless and water-soluble liquid that has a light smell of ammonia.

Flame Retardants

Brominated flame retardants, while banned by many governmental bodies due to their biologically persistent nature, are still relatively low cost and ubiquitously used where they are still legal. We are exposed to these compounds through upholstery fabrics, household furniture, and electronics. The chemicals are known to migrate out of these products; in time they become particulate matter in household dust and then are inhaled by humans. The European Union has put tight restrictions on the use of these compounds because of their endocrine-disrupting capabilities—known to negatively affect the development of reproductive organs, disrupt thyroid function and neurological development, and correlate to learning and behavior issues in children. In 2003 California issued a legislative ban on flame retardants, specifically focused on the use of polybrominated diphenyl ethers (PBDEs), and since that time a dozen states have instituted similar bans.[42] Before the ban took place, studies had shown that PBDEs were present in the breast milk of 100 percent of the sample population of new mothers that were tested; 30 percent of those tested had very high levels. Through continued biomonitoring studies conducted in 2009 and 2012—years after the ban took effect—it was shown that sixty-six first-time mothers tested at the Santa Rosa Women's Health and Birth Center had significantly lowered levels of PBDEs compared with eighty-two first-time mothers who were tested at the center in 2003 and 2006 (before the ban). This lowering of breast milk toxicity illustrates the positive impact that government policies can have on our health and the health of our children. The study also illuminates that we are capable of cleaning up our material culture and improving our collective quality of life if we decide it is important enough to us and if we are provided enough information to make informed decisions.

Antimony Trioxide

Antimony trioxide is a compound used as a catalyst in the production of polyester fiber and is often used in combination with brominated flame retardants. Antimony trioxide was found in thirty-six of the thirty-six children's articles of clothing tested by independent labs in Greenpeace's 2009 report *A Little Story About the Monsters in Your Closet*. Within the European Union's *Ecolabel Regulation*, antimony trioxide is limited to 260 mg/kg, although several items tested by the independent labs contained quantities well above the allowable levels—up to 293 mg/kg. Antimony trioxide is listed as a "possible carcinogen" to humans and is known to cause dermatitis and irritate the respiratory tract.[43] When rats were exposed to the substance, it was found to concentrate in their thyroids and gastrointestinal tracts; lower concentrations were found in the spleen, kidney, heart, bone, muscle, liver, and lungs.[44]

Fibershed Affiliates

The Fibershed Affiliate Program supports an international grassroots network that promotes the development of regional fiber systems communities, including economic and non-economic growth, in the form of building relationships and new global networks. In addition, a constantly updated version can be found on the Fibershed website: www.fibershed.com/affiliate-directory.

USA

Acadiana Fibershed

CONTACT: Sharon Donnan

EMAIL: acadianbrowncotton@gmail.com

WEBSITE: www.facebook.com/AcadianBrownCotton

STRATEGIC GEOGRAPHY: Southwest Louisiana: 13 parishes in Acadiana

MISSION STATEMENT: Field to Fashion in Acadiana: Careful and controlled selection, saving and replanting of Acadian Brown cotton is absolutely necessary for survival of this variety of *Gossypium hirsutum*. The preferred traits of this New World cotton were cultivated by humans for several millennia. Acadians and their descendants in Louisiana continued this selection regimen for over 200 years. We are proud to announce the University of Louisiana, Lafayette, Cade Research Farm, and Seed Bank are now growing our Acadian Brown Cotton heirloom seed.

Austin Fibershed

CONTACT: Weavers & Spinners Society of Austin; Inga Marie Carmel, Becky Johnson

EMAIL: atxfibershed@gmail.com

WEBSITE: www.atxfibershed.org

STRATEGIC GEOGRAPHY: Texas (entire state)

MISSION STATEMENT: A study group of the Weavers & Spinners Society of Austin, the Austin Fibershed researches the impact and viability of creating handmade wardrobes and household items from locally sourced fibers, dye, plants, and labor. As part of the journey, they are creating a wardrobe and a selection of home fashions from the locally sourced materials.

Chenango Fibershed

CONTACT: Kathryn Wojciechowski

EMAIL: chenangowoolworks@yahoo.com

STRATEGIC GEOGRAPHY: 175-mile radius from South Otselic, NY

Chesapeake Fibershed

CONTACT: Marian Bruno, Martha Polkey, Gretchen Frederick

EMAIL: chesapeakefibershed@gmail.com

STRATEGIC GEOGRAPHY: Approximately 120-mile radius from Richmond, VA

Connecticut Fibershed

CONTACT: Alisa Mierzejewski

EMAIL: alisamierzejewski@gmail.com

STRATEGIC GEOGRAPHY: 200-mile radius from Canterbury, CT (entire state)

Feather River Fibershed

CONTACT: Lorraine Nielsen

WEBSITES: www.featherriverfibershed.weebly.com; www.facebook.com /featherriverfibershed

STRATEGIC GEOGRAPHY: Plumas, Sierra, and Lassen Counties, CA

MISSION STATEMENT: The Feather River Fibershed is a regional organization encouraging a conscientious community that supports and connects local fiber producers, artisans, and consumers. We heighten public awareness of sustainable textile practices through education, highlighting the abundance of our rural mountain landscape—the plant, animal, and human resources that grace our region. Our focus includes revitalizing the natural fiber and dye heritage and creative skills that have become lost in the quest for speed, convenience, and profit in the broader textile industry. We envision a thriving future through the synergy of land stewardship, artisan creativity, and informed consumers.

Fibershed Carolinas

CONTACT: Stephanie Ondo
EMAIL: FibershedCarolinas@gmail.com
WEBSITES: facebook.com/FibershedCarolinas; sloyarnandfiber.com
INSTAGRAM: @fibershedcarolinas
STRATEGIC GEOGRAPHY: 300-mile radius from Charleston, SC

Fibershed of the Greater Northwest

CONTACT: Karmel Ackerman
EMAIL: info@fibersoftheworld.com
WEBSITE: www.facebook.com/Fibershed-of-the-Greater-Northwest
  -1033973589962632
STRATEGIC GEOGRAPHY: 300-mile radius from Seattle, WA

Great Lakes Fibershed

CONTACT: Amanda Cinco-Hoyt, Lauren Mathieson, Michelle McCoskey
EMAIL: colorwheelmichigan@gmail.com
STRATEGIC GEOGRAPHY: 250-mile radius from Detroit, MI
MISSION STATEMENT: The Great Lakes Fibershed focuses on a radius
  of 250 miles from Detroit, Michigan, and includes portions of
  Canada, Indiana, Illinois, Ohio, New York, and Pennsylvania. Our
  Fibershed welcomes community involvement between fiber
  producers, mills, dyers, artisans, and enthusiasts in an effort to
  unify local resources and promote environmentally sustainable
  practices. Currently guided by three core members, we look
  forward to expanding our network of supporters and contributors
  and invite all who are interested to share their ideas, knowledge,
  and skills as our Fibershed organically takes shape.

High Desert Fibershed of California

CONTACT: Bonnie Chase, Warner Mountain Weavers
EMAIL: warnermtnweavers@citlink.net
STRATEGIC GEOGRAPHY: Northeast CA, southern OR, Reno NV

Inland Empire of Southern California Fibershed

CONTACT: Mary Hurley
EMAIL: finefleeces@aol.com
STRATEGIC GEOGRAPHY: 75-mile radius from Redlands, CA

Kaw Point Fibershed

CONTACT: Jamie Root

WEBSITES: kawpointfibershed@gmail.com; kawpointfibershed.com; www.facebook.com/kawpointfibershed

STRATEGIC GEOGRAPHY: 150-mile radius from Kansas City, MO (eastern Kansas and western Missouri surrounding the confluence of the Kansas and Missouri Rivers in Kansas City)

LocalFiber NY Fibershed

CONTACT: Dana Havas

WEBSITE: localfiber.org

INSTAGRAM: @localfiber.ny

STRATEGIC GEOGRAPHY: 150-mile radius from Ithaca, NY

MISSION STATEMENT: LocalFiber works with small flock fiber producers in the Finger Lakes/Central region of New York. Small flock fiber producers throughout the northeastern United States face numerous challenges in getting their goods to market, from shearing to accessing the market.

LocalFiber is a group of small flock fiber producers coming together to collectively identify and address the challenges they face.

LocalFiber is a community made up of the farmers and makers that value locally produced animal fibers. Together we are a part of a strong local economy.

Maine Fibershed

CONTACT: Marcia MacDonald; Rachel Bingham Kessler

WEBSITE: facebook.com/mainefibershed

INSTAGRAM: @mainefibershed

STRATEGIC GEOGRAPHY: Maine (entire state)

Mountains & Plains Fibershed

CONTACT: Anne-Marie Chmielewski, Karey Grant

EMAIL: localinfo@mountainsplainsfibershed.com

WEBSITE: mountainsandplainsfibershed.org

INSTAGRAM: @fibershedmountainsplains

STRATEGIC GEOGRAPHY: 300-mile radius from Golden, CO

MISSION STATEMENT: We foster collaboration among textile artists, designers, fiber farmers, processing mills, suppliers, and retail businesses in and around Colorado.

Monterey Bay Fibershed

CONTACT: Trish Sparling

EMAIL: tsparling@sbcglobal.net

WEBSITE: fibershedmontereybay.com

STRATEGIC GEOGRAPHY: Royal Oaks, CA (60-mile radius from Monterey Bay)

NJ Fibershed

CONTACT: Anne Choi

EMAIL: njfibershed@gmail.com

WEBSITE: www.njfibershed.com

STRATEGIC GEOGRAPHY: New Jersey (entire state)

MISSION STATEMENT: NJ Fibershed is working to create a network among independent fiber farmers raising small flocks of sheep, alpacas, goats, and other fiber animals in the state of New Jersey. Our goal is to promote face-to-face relationships between fiber animals, farmers, and fiber artists within our community. NJ Fibershed provides educational programs focused on small-scale fiber farming, such as animal care and fiber processing, and we organize community outreach events for current and aspiring fiber farmers.

Northern Minnesota Fibershed

CONTACT: Keila McCracken

EMAIL: northernmnfibershed@gmail.com

WEBSITE: www.northernminnesotafibershed.com

STRATEGIC GEOGRAPHY: 200-mile radius from Bemidji, MN

Pacific Northwest Fibershed

CONTACT: Shannon M Welsh

EMAIL: pnwfibershed@gmail.com

WEBSITE: www.pacific-northwest-fibershed.com

INSTAGRAM: @pnw_fibershed

STRATEGIC GEOGRAPHY: 300-mile radius from Portland, OR

MISSION STATEMENT: Pacific Northwest Fibershed envisions the emergence of a bioregional textile community that will support soil-to-soil processes in fiber, textile, and garment production. Our natural textile resource base is defined by the geographical landscape of a 300-mile radius from Portland, Oregon. After thoroughly researching the successes and mistakes of the Pacific Northwest's historic fiber industry, we have identified our community's need for natural fibers and its proven ability to grow a quality crop. Together we will inspire our regenerative textile systems to participate in the revitalization of the Pacific Northwest textile industry.

Piedmont Fibershed

CONTACT: Courtney Lockemer
EMAIL: piedmontfibershed@gmail.com
WEBSITE: facebook.com/piedmontfibershed
STRATEGIC GEOGRAPHY: 200-mile radius from Durham, NC

Rust Belt Fibershed

CONTACT: Jessalyn Boeke, Sarah Pottle
EMAIL: rustbeltfibershed@gmail.com
INSTAGRAM: @rustbeltfibershed
WEBSITE: rustbeltfibershed.com
STRATEGIC GEOGRAPHY: 250-mile radius from Cleveland, OH
MISSION STATEMENT: The Rust Belt Fibershed consists of a 250-mile radius outside of Cleveland, Ohio, including parts of Michigan, southern Ohio, western Pennsylvania, and western New York. We are building a community that collaboratively supports locally grown textiles. We aspire to connect everyone in all parts of our local fiber system: farmers, fiber processors—from large mills to home spinners, weavers, dyers, and fiber artists—to designers, shop owners, consumers, and fiber enthusiasts. Through this project, we hope to foster friendship, creativity, and a greater respect for our environment, as well as an understanding of the impact we share in our corner of the world.

San Diego Fibershed

CONTACT: Susan Plack
WEBSITE: facebook.com/SanDiegoFibershed
STRATEGIC GEOGRAPHY: Greater San Diego, CA

Sligo Creek Fibershed

CONTACT: Kerstin Zurbrigg, Ava Stebbins
STRATEGIC GEOGRAPHY: 100-mile radius from Silver Spring, MD

Southeastern New England Fibershed

CONTACT: Sarah Kelley, Karen Schwalbe, Amy DuFault
WEBSITE: senefibershed.org
INSTAGRAM: @senewenglandfibershed
EMAIL: fiber@semaponline.org
STRATEGIC GEOGRAPHY: Massachusetts and Rhode Island
MISSION STATEMENT: The Southeastern New England Fibershed comprises Massachusetts and Rhode Island. Our mission is to

create a regional fibershed that unites fiber to finished product. From farmer to processor, from financing to cut and sew, we are connecting the dots of the supply chain to bring production back to reinvigorate a once-thriving New England textile industry. We base our geographic radius on the historical textile processing centers of New Bedford, Massachusetts, and Providence, Rhode Island, both of which have extensive remaining infrastructure, and cover both states with an approximately 100-mile radius.

Southern Appalachia Fibershed

CONTACT: Kacie Lynn
WEBSITE: safibershed.com
INSTAGRAM: @southernappalachiafibershed
FACEBOOK: Southern Appalachia Fibershed
STRATEGIC GEOGRAPHY: 200-mile radius from Knoxville, TN
MISSION STATEMENT: The Southern Appalachia Fibershed consists of a 200-mile radius outside of Knoxville, Tennessee, and includes parts of north Alabama, Georgia, western North Carolina, and Kentucky. We began as a way of building community and connecting the entire supply chain from farmer to consumer. Through this collaborative network we look forward to strengthening our local economy and playing a supportive role in raising awareness about our region's variety of local textile resources.

Southern California Fibershed

CONTACT: Lesley Roberts
EMAIL: hello@socalfibershed.org
WEBSITE: www.socalfibershed.org
STRATEGIC GEOGRAPHY: The region south of San Luis Obispo to San Diego, from the Pacific Coast to the Arizona border
MISSION STATEMENT: Our Fibershed encourages community building and connections between textile artists, designers, fiber farmers, processing mills, suppliers, and retail businesses. We encourage a transparency that will empower wearers, farmers, artisans, and consumers to engage with and understand the soil-to-skin paradigm. We seek to deepen the sustainable textile conversation by directly connecting the wearer to the biological context of their wardrobe. Southern California is an urban center dense with talented designers, makers, and artisans eager to engage with and understand alternative processes. The So Cal Fibershed aims to generate awareness and teach the necessary skills to build and

sustain a thriving textile culture that functions hand-in-hand with principles of ecological balance, local economies, and regional organic agriculture.

Textile Lab Fibershed

CONTACT: Laura Sansone
EMAIL: nytextilelab@gmail.com
WEBSITE: www.newyorktextilelab.com
INSTAGRAM: @nytextilelab
STRATEGIC GEOGRAPHY: 300-mile radius from New York City, NY
MISSION STATEMENT: We bring New York designers together with fiber producers and mills to help to grow an economically diverse, statewide textile supply chain. We aim to rebuild regional textile production, which adds value to our communities and the environment.

Designers recognize the challenges that plague large manufacturing systems, such as environmental violations leading to climate change, human rights abuses in factories, and the continuing centralization of economic power. Small-scale approaches to manufacturing allow control throughout production systems so greater attention is paid to the well being of people and the environment. Our goal is to create distributed networks of small businesses that include fiber farms, mills, and manufacturers, which enable the growth of a resilient textile economy in New York State.

Three Rivers Fibershed

CONTACT: Maddy Bartsch
EMAIL: threeriversfibershed@gmail.com
WEBSITE: www.threeriversfibershed.com
INSTAGRAM: @threeriversfibershed
STRATEGIC GEOGRAPHY: 175-mile radius from Minneapolis, MN
MISSION STATEMENT: The Three Rivers Fibershed is working to develop regional fiber systems that build soil and protect the health of our biosphere here in Minnesota. Working with a strategic geography centered at the Textile Center of Minnesota and extending out in a 175-mile radius, we include portions of Minnesota, Wisconsin, Iowa, and South Dakota. The Three Rivers Fibershed aims to come alongside our community to enhance our local fiber systems, finding what needs to exist to help uplift all members of the fiber community. We seek to provide opportunities for connection among farmers and mills, artists and makers, consumers and everyone in between, strengthening our local fiber communities.

Vashon Fibershed

CONTACT: Laurel Boyajian, Emily MacRae
EMAIL: vashonfibershed@gmail.com
STRATEGIC GEOGRAPHY: 100-mile radius from Vashon Island, WA
MISSION STATEMENT: Working within a 100-mile radius of Vashon Island, our Fibershed seeks to utilize local materials whenever possible, including animal and plant fibers, dye plants, and upcycled fabric. As an affiliate of the greater Fibershed nonprofit organization, we are connected to groups throughout the world that are dedicated to the soil-to-skin movement.

Western Mass Fibershed

CONTACT: Nur Tiven, Michelle Parrish, Lisa Bertoldi
WEBSITE: westernmassfibershed.org
INSTAGRAM: @westernmassfibershed
EMAIL: WesternMassFibershed@gmail.com
STRATEGIC GEOGRAPHY: Connecticut River Valley & Berkshires (100-mile radius)
MISSION STATEMENT: The Western Massachusetts Fibershed extends from the Connecticut River Valley to the Berkshires in Massachusetts and includes parts of southern Vermont. We are organizing to support the creation of locally sourced sustainable clothing and textiles in western Massachusetts. We are farmers, weavers, spinners, dyers, designers, knitters, and sewists working together to grow and make our own clothes from our regional Fibershed.

International

Fibreshed Melbourne

CONTACT: Rachel Bucknall, Nicki Colls
EMAIL: letsyarn@fibreshedmelbourne.com
WEBSITE: www.fibreshedmelbourne.com
INSTAGRAM: @fibreshedmelbourne
MISSION STATEMENT: Fibreshed Melbourne supports our local economy to access locally produced, sustainable textiles. We want our clothes to benefit the land and people in and around Melbourne.

We advocate for a transparent, accessible, and sustainable local textile network from its beginnings at the farm to its return to the soil at the end of its life. We do this by expanding carbon farming opportunities, rebuilding the foundations of regional manufacturing, and connecting end-users to their local supply chain in a personal way.

Fibreshed Melbourne is the only organisation in Australia that takes a cross-sector view of the textile ecosystem. Our holistic, relationship driven approach delivers desirable, viable and feasible solutions for the entire community.

Midt-Norge Fibershed

CONTACT: Årolilja Jørgensrud

STRATEGIC GEOGRAPHY: The counties of Trøndlag and Møre and Romsdal in Norway; a 125-mile radius from the city of Trondheim

Northern England Fibreshed

CONTACT: Claudia Quayle

EMAIL: claudia@thetextillery.co.uk

STRATEGIC GEOGRAPHY: Northumberland, Cumbria, Lake District, County Durham, Tyne and Wear, North Yorkshire

One Year One Outfit Fibreshed

CONTACT: Nicki Colls

EMAIL: thisismoonlightsewing@gmail.com

WEBSITES: thisismoonlight.com/1year1outfit; www.facebook.com /1year1outfit

MISSION STATEMENT: One Year One Outfit is a challenge to make a locally sourced outfit in a year. Anyone interested in garment making is welcome to join in. Most participants record their findings through social media and use the tag #1year1outfit to keep in touch with the group.

Pembina Fibreshed

CONTACT: Margaret Brook, Anna Hunter

EMAIL: connect@pembinafibreshed.com

WEBSITE: www.pembinafibreshed.com

STRATEGIC GEOGRAPHY: 200-mile radius from Pembina River Valley, Manitoba, Canada

MISSION STATEMENT: We work to educate, advocate for, and support fibre producers, manufacturers, and dyers within the Pembina Fibreshed. Our goal is to support local farmers, dyers, spinners, and designers to produce more local textiles and to get that in the hands of local knitters, spinners, weavers, and sewists. We want to see a thriving Manitoba fibre economy.

Prakriti Fibershed

EMAIL: n.chopra@oshadi.in

STRATEGIC GEOGRAPHY: 8 acres within Erode, Tamil Nadu

MISSION STATEMENT: Prakriti Fibershed is aiming to revive and sustain traditional and sustainable cotton farming practices and enhance the livelihood of small-holder cotton farmers. Overall the goal of the project is to create self-sufficient communities in remote villages, who along with organic farming, can be directly involved in various steps of the cotton supply chain through to the final fabric and garment making. Beginning with a 5-acre model project to inspire more local/regional farmers to adopt organic farming practices, fostering employment and empowerment among the village communities and ensuring a fair income distribution throughout the supply, the project will also aim to address diverse sustainable farming needs including waste management (through composting), soil water conservation and rainwater harvesting, carbon farming, and residue-free and smart agricultural practices.

Salt Spring Island Fibreshed

CONTACT: Terri Bibby

WEBSITES: www.saltspringfibreshed.com; www.facebook.com /saltspringislandfibreshed

STRATEGIC GEOGRAPHY: Salt Spring Island and Southern Gulf Islands, Canada

MISSION STATEMENT: Like a watershed—where all the water from an area feeds into a local body of water, a fibreshed is an area where all the fibre, producers, products, and people work together. I hope to connect fibre farms, dye plants, textile mills, fibre artists, designers, weavers, felters, spinners, seamstresses, knitters, and others to produce items made from locally grown, milled, designed, and sewn fibres. Our focus is on Salt Spring Island and the Southern Gulf Islands of British Columbia, Canada.

South West England Fibreshed

CONTACT: Emma Jane Hague

EMAIL: emma@bristoltextilequarter.co.uk

WEBSITES: http://bristoltextilequarter.co.uk; www.bristolcloth.co.uk /#concept; www.facebook.com/workingwool

STRATEGIC GEOGRAPHY: Bristol, UK

Sunshine Coast BC Fibreshed

CONTACT: Merrily Corder, Lynda Daniells, Ann Harmer

WEBSITES: www.sunshinecoastfibreshed.ca; facebook.com
/sunshinecoastfibreshed

STRATEGIC GEOGRAPHY: 100 miles of coastline from Langdale to Lund,
BC, Canada

MISSION STATEMENT: Our mission is to decrease our environmental
footprint by encouraging and supporting awareness within the
community of the many uses of local animal and plant fibre and to
connect local fibre farmers with local artisans keeping within a
"slow fashion" model. We have a website, blog, directory of artisans,
producers, and vendors and promote Fibreshed activities through
social media such as Facebook. We host events, organize classes
and workshops, and attend many local community festivals and
gatherings to promote the message and work of Fibreshed.

Upper Canada Fibreshed

CONTACT: Jennifer Osborn, Becky Porlier

WEBSITES: www.uppercanadafibreshed.ca; www.facebook.com
//UpperCanadaFibreshed

INSTAGRAM: @uppercanadafibreshed

STRATEGIC GEOGRAPHY: 250-mile radius from Toronto, Canada

MISSION STATEMENT: The Upper Canada Fibreshed connects Ontario
fibre farmers, processors, artisans, and consumers collaborating for
a bioregional textile supply chain, within 250 miles of Toronto. The
organization acts as a hub for sharing skills and resources, empowering the creation of soil-to-soil value chains and inspiring the
shifts in perception necessary to appreciate the ecological resources
of our landscape.

Vancouver Fibreshed

STRATEGIC GEOGRAPHY: 150-mile radius from Vancouver, BC, Canada

Vancouver Island Fibreshed

CONTACT: Sara Mullin

STRATEGIC GEOGRAPHY: 150 miles of Comox Valley, Victoria to Port
Hardy, Canada

Resources

Hands-On Educational Opportunities

Berea College (KY) (for first-time college-bound Kentucky youth)
Hand Made Studio (CA)
John C. Campbell Folk School (NC)
Marshfield School of Weaving (VT)
North House Folk School (MN)
Shakerag Folk School (TN)
Textile Arts Center (NYC)

Local Guilds

ALL REGIONS

Crochet Guild of America
Crochet guilds
CONTACT NUMBER: (740) 452-4541
WEBSITE: www.crochet.org

Handweavers Guild of America Inc.
Spinning, weaving, dyeing guilds
EMAIL: hga@weavespindye.org
WEBSITE: www.weavespindye.org

Interweave Spinning Guild Directory
List of spinning guilds in all 50 states
WEBSITE: www.interweave.com/spinning
-guilds-directory

The Knitting Guild Association
Knitting guilds
EMAIL: info@tkga.org
WEBSITE: tkga.org

Natural Dyes International
Natural dye organization
EMAIL: conference@naturaldyes.org
WEBSITE: www.naturaldyes.org

NORTHEAST

Baa Baa Black Sheep Knitting Guild
Knitting; Stoneham, MA
www.facebook.com/BaaBaaBlack
SheepKnittingGuild

Black Sheep Handspinners Guild
Spinning; Lansing, NY
www.blacksheephandspinnersguild.org

Boston Area Spinners and Dyers
Spinning, dyeing; Boston, MA
www.basdspinning.org

Boston Weavers Guild
Weaving; Andover, MA
www.weaversguildofboston.org

Carroll County Knitters' Guild
Knitting; Carroll County, MD
www.ravelry.com/groups/carroll-county
-knitters-guild

Chautauqua County Weavers' Guild
Weaving; Westfield, NY
mafafiber.org/guilds

Chesapeake Weavers
Weaving; Greater Annapolis area, MD
mafafiber.org/guilds

Finger Lakes Machine Knitting
Machine knitting; Cortland, NY
www.fingerlakesknitting.com/home

Genesee Valley Handspinners
Spinning; Rochester, NY
www.gvhg.org

Golden Fleece Spinners' Society
Spinning; upstate NY
www.goldenfleecespinnerssociety.org

Handweavers Guild of Connecticut
Weaving, spinning; Cheshire, CT
www.handweaversguildofct.org

Handweavers of Bucks County
Weaving; Washington Crossing, PA
hwbcguild.org

Harmony Weavers Guild
Weaving; Delaware, New Jersey, Maryland,
    and Pennsylvania
harmonyweaversguild.org

Lancaster Spinners and Weavers Guild
Spinning, weaving; Leola, PA
www.lancspinweaveguild.webs.com

Lighthouse Knitting Guild of Maine
Knitting; southern Maine
www.facebook.com
    /lighthouse.knitters

Long Island Machine Knitters Guild
Machine knitting; Long Island, NY
www.limkg.com

Maine Spinners Registry
Spinning; Augusta, ME
www.mainespinnersregistry.org

MidAtlantic Fiber Association
Fiber arts; Greater Mid-Atlantic region
www.mafafiber.williamsburgspinweave
    .org/wordpress

New Hampshire Weavers Guild
Weaving; Concord, NH
www.nhweaversguild.org

Northeast Feltmakers Guild
Felting; New England and other
    northeastern states
northeastfeltmakersguild.org

**Northwest Pennsylvania Spinners and
Weavers Guild**
Spinning, weaving; northwest Pennsylvania
www.npswg.org

Nutmeg Spinners Guild
Spinning; West Hartford, CT (statewide guild)
www.nutmegspinners.org

Palisades Guild of Spinners and Weavers
Spinning, weaving; Northern NJ/Rockland
    County, NY
www.mafafiber.org/palisades
    /index.html

Rhode Island Spinners Guild
Spinning; Rhode Island
www.facebook.com/Rhode-Island
    -Spinners-Guild-Page-1374812999499152

South Cumberland Fiber Guild
Spinning, weaving, dyeing, knitting, crocheting;
    south-central Pennsylvania
www.southcumberlandfiberguild.com

**Southern Maryland Sticks and Stitches
Knitting Guild**
Knitting and crocheting; southern Maryland
www.ravelry.com/groups/so-md
    -sticks--stitches

Spin City—NY
Spinning; New York, NY
www.meetup.com/spinners-87

Spinning Study Group of Long Island
Spinning; Long Island, NY
www.lispinning.com

**Susquehanna Valley Spinners and
Weavers Guild**
Spinning, weaving; Lewisburg, PA
www.svswg.org

Syracuse Weavers Guild
Weaving; Syracuse, NY
www.syracuseweaversguild.org

Twist O' Wool Guild
Spinning, dyeing, weaving; Middlebury, VT
twistowoolguild.weebly.com

Ulster County Handspinners Guild
Spinning; Lake Katrine, NY
ulsterspinners.wordpress.com

Weavers Guild of Buffalo
Weaving; Amherst, NY
weaversguildofbuffalo.org

Weavers Guild of Rochester
Weaving; Rochester, NY
www.weaversguildofrochester.org
    /index.php

MIDWEST

Ann Arbor Fiberarts Guild
Felting, spinning, weaving, dyeing, knitting,
    crocheting; Ann Arbor, MI
annarborfiberarts.org

Arrow Rock Weavers and Spinners Guild
Weaving, spinning, dyeing, knitting,
    crocheting; Boonville, MO
www.arrowrockweavers.org

Bishop Hill Fiber Guild
Spinning, knitting, crocheting; Chillicothe, IL
bishophillfiberguild.org

Central Kansas Knitters Guild
Knitting; Central Kansas
centralksknittersguild.com

Central Ohio Weavers Guild
Weaving, spinning, dyeing; Centerburg, OH
www.cowg.org

**Champaign Urbana Spinners and
Weavers Guild**
Spinning and weaving; Champaign, IL
centralillinoisfiberguild.org

Columbia Weavers & Spinners Guild
Weaving, spinning; Columbia, MO
cwsg.missouri.org

The Cultured Purls
Knitting; McHenry County, IL
www.culturedpurls.com

**Des Moines Weavers and
Spinners Guild**
Weaving, spinning; Des Moines, IA
www.desmoinesweaversguild.org

Duluth Fiber Handcrafters Guild
Spinning, weaving, dyeing, knitting, felting;
    Duluth, MN
duluthfiberhandcrafters.org

**Duneland Weavers Guild of
Northwest Indiana**
Weaving; Chesterton, IN
www.dunelandweaversguild.org/dwg
    /index.htm

**Fiber Arts Guild of the
Red River Valley**
Spinning, weaving, dyeing, knitting, felting;
    Grand Forks, ND
fiberartsguildrrv.weebly.com

Fiber Guild of Kansas City
Weaving, spinning, knitting; Kansas City, MO
kansascityfiberguild.org

Fiber of Our Being Felting Guild
Felting; Toledo, OH
www.therovingartist.com/fob
    -felting-guild.html

Fox Valley Spinning Guild
Spinning, weaving, felting, knitting,
    crocheting; Appleton, WI
www.foxvalleyspinningguild.com

Handweavers Guild of Peoria
Weaving, spinning; Peoria, IL
peoriaweavers.wordpress.com

Illinois Prairie Weavers Guild
Weaving; Lombard, IL
www.illinoisprairieweavers.org

InterKnits Machine Knitters Guild
Machine knitting; Glen Ellyn, IL
www.interknitmk.org

Kansas Alliance of Weavers & Spinners
Weaving, spinning; Emporia, KS
www.kansasweaversspinners.com

Machine Knitters Guild of Minnesota
Machine knitting; Brooklyn Center, MN
www.facebook.com/mkgmn

Madison Knitters Guild
Knitting; Madison, WI
www.madisonknittersguild.org

Michigan League of Handweavers
Weaving; Kalamazoo, MI
www.mlhguild.org

Midwest Machine Knitters Collaborative
Machine knitting; Minneapolis, MN
www.midwestmachineknitters.org

Midwest Weavers Association
Weaving; entire Midwest
www.midwestweavers.org/index.htm

Minnesota Feltmakers Guild
Felting; Minneapolis, MN
www.facebook.com/Minnesota
    FeltmakersGuild

Minnesota Knitters Guild
Knitting; Minneapolis, MN
knitters.org

Northcoast Knitting Guild
Knitting; Mayfield Heights, OH
www.northcoastknitting.org

Northeast Iowa Weavers and Spinners
Weaving, spinning; Waterloo, IA
neiwsguild.wordpress.com

Omaha Weavers and Spinners Guild
Spinning, weaving; Omaha, NE
www.owsguild.com

Shuttlecraft Guild
Weaving; Sheboygan County, WI
shuttlecraftguild.org

South Central Michigan Knitting Guild
Knitting; south-central Michigan
scmkguild.webs.com

Spinners and Weavers of Indiana Fiber and Textiles (SWIFT)
Spinning, weaving; Indiana (statewide)
swiftindiana.org

Topeka Weavers Guild
Weaving; Topeka, KS
topekahandweaversandspinners
    .wordpress.com

Wabash Weavers Guild
Weaving; Lafayette, IN
wabashweaversguild.smugmug.com

Weavers Guild of Greater Cincinnati
Weaving, spinning, felting, dyeing, knitting;
    Cincinnati, OH
weaversguildcincinnati.org

Weavers Guild of Greater Kansas City
Weaving, spinning, dyeing; Kansas City, MO
www.kcweaversguild.org

Weavers Guild of Minnesota
Weaving, spinning, dyeing; Minneapolis, MN
www.weaversguildmn.org

Weavers Guild of St. Louis
Weaving, spinning, dyeing; St. Louis, MO
www.weaversguildstl.org

Weaving Indiana
Weaving; Indianapolis, IN
www.weavingindiana.org

Western Reserve Knitting Guild
Knitting; Canfield, OH
www.facebook.com
    /westernreserveknittingguild

Western Reserve Weavers & Spinners
Weaving, spinning; Aurora, OH
www.wrspinweave.org

Wichita Weavers, Spinners & Dyers Guild
Weaving, spinning, dyeing; Derby, KS
wichitaweavers.org

Wisconsin Handweavers Inc.
Weaving; Milwaukee, WI
wihandweavers.org

Woodstock Weavers Guild
Weaving; Woodstock, IL
sites.google.com/site/woodstock
    weaversguild2

Zumbro River Fiber Arts Guild
Spinning, weaving, dyeing, knitting, felting;
    near Rochester, MN
zumbroriverfiberartsguild.com

SOUTH

Bay Area Knitting Guild
Knitting; Tampa Bay, FL
www.bayareakg.com

Carolinas Machine Knitters Guild
Machine knitting; North and South Carolina
www.carolinasmkg.com

Central Arkansas Spinning Guild
Spinning; Little Rock, AR
www.ravelry.com/groups/central
    -arkansas-spinning-guild

Chimneyville Weavers and Spinners Guild
Weaving, spinning; Ridgeland, MS
www.cvillewsg.com

Clotho's Handspinners
Spinning; Richmond, VA
www.clothos.org

Contemporary Handweavers of Houston
Weaving, spinning, dyeing; Houston, TX
www.weavehouston.org

Contemporary Handweavers of Texas
Weaving; Texas (statewide)
www.weavetexas.org

CSRA Knitters Guild
Knitting; Central Savannah River area, SC
www.facebook.com/CsraKnittersGuild

Dallas Handweavers and Spinners Guild
Weaving, spinning; Dallas, TX
www.dallasweavers.org

DFW Machine Knitters Guild
Machine knitting; Dallas–Fort Worth, TX
www.dfwmachineknitters.com

El Paso Knitting Guild
Knitting; El Paso, TX
www.elpasoknittingguild.com

Fiber Artists of Oklahoma
Weaving, spinning, felting, knitting,
    crocheting, dyeing; Oklahoma City, OK
fiberartistsok.org

Foothills Spinners and Weavers
Spinning, weaving; Reston, VA
www.foothillsfiber.com

Greater Birmingham Fiber Guild
Weaving, spinning, dyeing, felting, knitting,
    crocheting; Birmingham, AL
greaterbirminghamfiberguild
    .blogspot.com

Florida Tropical Weavers
Spinning, weaving; statewide guild, Florida
www.ftwg.org

Houston Area Machine Knitters
Machine knitting; Houston, TX
www.houstonmk.com

The Knit at Night Guild
Knitting, crocheting, spinning, weaving;
    Greater Houston, TX
www.knitatnight.org

Knitting Artists of Northern Virginia
Knitting; northern Virginia
kanv.org

K.O.M. Rivers Knitting Guild
Knitting; Parkersburg, WV
www.ravelry.com/groups/KOM-rivers
   -Knitting-guild

Manasota Weavers Guild
Weaving, spinning, dyeing; Sarasota, FL
www.manasotaweaversguild.com

Memphis Guild of Handloom Weavers
Weaving; Memphis, TN
www.mghw.org

Metro Atlanta Machine Knitters
Machine knitting; Atlanta, GA
www.ravelry.com/groups/browse/show
   /metro-atl-machine-knitters

North Georgia Knitting Guild
Knitting; north Georgia
northgeorgiaknittingguild.com

Northwest Arkansas Handweavers Guild
Weaving; Springdale, AR
nwahandweaversguild.com

Ol' North State Knitting Guild
Knitting; Raleigh, NC
www.meetup.com/Ol-North-State
   -Knitting-Guild

Palmetto Fiber Arts Guild
Weaving, spinning, dyeing, felting, knitting,
   crocheting; Charleston, SC
www.palmettofiberartsguild.blogspot.com

Peachtree Handspinners Guild
Spinning; Stone Mountain, GA
www.peachtreehandspinnersguild.org

Piedmont Fiber Guild
Weaving, spinning, dyeing, felting, knitting;
   Gastonia/Charlotte, NC
www.piedmontfiberguild.org

Potomac Fiber Arts Guild
Spinning, weaving, dyeing, felting, knitting;
   Washington, DC
www.potomacfiberartsguild.org

Roanoke Valley Knitting Guild
Knitting; Roanoke, VA
www.thinkingknitter.com/Roanoke
   ValleyKnittingGuild.htm

Tangled Threads
Weaving, spinning, knitting, crocheting,
   machine knitting, felting; CSRA, GA
www.facebook.com/tangledthreadscsra

Triangle Machine Knitters
Machine knitting; Raleigh, NC
sites.google.com/site/triangle
   machineknitters

Triangle Weavers
Weaving, spinning, dyeing, felting, knitting;
   Chapel Hill, NC
www.triangleweavers.org

Weavers and Spinners Society of Austin
Weaving, spinning; Austin, TX
wssaustin.org

Weavers Guild of Greater Baltimore
Weaving, spinning, dyeing; Baltimore, MD
www.wggb.org

WEST

Anchorage Weavers and Spinners Guild
Weaving, spinning; Anchorage, AK
www.anchorageweavespin.org

Arizona Desert Weavers and Spinners Guild
Weaving, spinning; Phoenix, AZ
www.adwsg.org

**Arizona Federation of Weavers
and Spinners**
Weaving, spinning; statewide (the website lists
   guilds all across Arizona)
www.azfed.org/azfed

Association of Northwest Weavers' Guilds
Weaving; entire Northwest
northwestweavers.org

Bisbee Fiber Arts Guild
Weaving, spinning; Bisbee, AZ
www.bisbeefiberarts.com

Cactus Needles Knitting Guild
Knitting; Phoenix, AZ
www.cactusneedlesknittingguild.com

Carmel Crafts Guild
Weaving, knitting, spinning, dyeing; Monterey
    County, CA
www.carmelcraftsguild.org

Central Coast Weavers
Weaving; Northern Santa Barbara / San Luis
    Obispo County, CA
www.centralcoastweavers.org

Desert Fiber Arts
Weaving, spinning; Richland, WA
www.desertfiberarts.org

Eastside Spinners Guild
Spinning; Issaquah, WA
groups.yahoo.com/neo/groups
    /eastsidespinners/info

El Segundo Slipt Stitchers
Knitting; El Segundo, CA
www.sliptstitchers.org

Eugene Weavers Guild
Weaving; Eugene, OR
www.eugeneweavers.com

Fairbanks Weavers and Spinners Guild
Weaving, spinning; Fairbanks, AK
www.fairbanksweavers.org

Fiber to Finish
Weaving, spinning, knitting, crocheting,
    felting; Valencia County, NM
www.fibertofinish.org

Foothill Fibers Guild
Spinning, weaving, knitting; Nevada City, CA
www.foothillfibersguild.org

Front Range Knitting Guild
Knitting; Northern Colorado
frkg.webs.com

Great Falls Spinners & Weavers Guild
Spinning, weaving; Great Falls, MT
www.greatfallsspinnersweaversguild
    .weebly.com

Greater Los Angeles Spinning Guild
Spinning; Los Angeles, CA
www.glasg.org

Handweavers Guild of Boulder
Weaving; Boulder, CO
www.handweaversofboulder.org

Hawaii Handweavers' Hui
Spinning, weaving, dyeing; Honolulu, HI
www.hawaiihandweavers.org

Job's Peak Fiber Arts
Weaving, spinning, dyeing, knitting,
    crocheting; Carson Valley, NV
www.jobspeakfiberarts.com

Knerdy Knitters of the SFV
Knitting; San Fernando Valley, CA
www.knerdyknitters.com

Las Aranas Spinners & Weavers Guild
Spinning, weaving; Albuquerque, NM
www.lasaranas.org

Las Tejadoras
Weaving, spinning, knitting, crocheting,
    felting, dyeing; Santa Fe, NM
www.lastejedoras.org

**Machine Knitters Guild of
San Diego**
Machine knitting; San Diego, CA
www.mkgsd.com

Machine Knitters Guild San Francisco Bay Area
Machine knitting; San Francisco, CA
www.mkgsfba.org

Mountain Spinners & Weavers Guild
Weaving, spinning; Prescott, AZ
www.mtnspinweave.org

North Olympic Shuttle and Spindle Guild
Weaving, spinning, dyeing, felting, knitting, crocheting; Northern Olympic Peninsula, WA
www.nossg.org

Northwest Regional Spinners Association
Spinning; Pacific Northwest
www.nwregionalspinners.org

Oregon Metropolitan Machine Knitters Guild
Machine knitting; Oregon City, OR
mmkg.freehostia.com

Pikes Peak Weavers Guild
Weaving, spinning, dyeing; Colorado Springs, CO
www.pikespeakweavers.org

Redwood Guild of Fiber Arts
Weaving, spinning; Northern California
www.redwoodgfa.org

Reno Fiber Guild
Weaving, spinning; Reno, NV
renofiberguild.blogspot.com

Rocky Mountain Weavers Guild
Weaving, dyeing; Denver, CO
www.rmweaversguild.org

Sacramento Area Machine Knitters Guild
Machine knitting; Sacramento, CA
machineknittingfun.blogspot.com

Salt Lake Knitting Guild
Knitting; Salt Lake City, UT
saltlakeknittingguild.org

San Diego Creative Weaver's Guild
Weaving; San Diego, CA
www.sdcwg.org

San Juan County Textile Guild
Dyeing, felting, crocheting, knitting, weaving, spinning; Puget Sound, WA
www.sjctextileguild.org

Seattle Weavers Guild
Weaving; Seattle, WA
seattleweaversguild.com

South Bay Machine Knitting Meetup Group
Machine knitting; San Jose, CA
www.meetup.com/machineknit-103

Southern California Handweavers Guild
Weaving; Harbor City, CA
www.schg.org

Spindles & Flyers Spinning Guild
Spinning; Berkeley / San Francisco Bay Area, CA
www.spindlesandflyers.org

Telarana Spinners and Weavers
Weaving, spinning; Mesa, AZ
www.telarana.org

Tigard Knitting Guild
Knitting; metropolitan area near Portland, OR
www.tigardknittingguild.org

Twisted Threads Fiber Guild
Weaving; Vernon & White Mountain region, AZ
www.azfed.org

Valley Fiber Arts Guild
Weaving, spinning, knitting; Mat-Su Valley, AK
www.valleyfiberarts.org

Weaving, Spinning and Fiber Arts Guild of Idaho Falls
Weaving, spinning, felting, dyeing, knitting; Idaho Falls, ID
www.srfiberartists.org

What the Knit! Guild
Knitting; Bakersfield, CA
what-the-knit.org

Campaigns & More Websites That Provide Education

Care What You Wear: www.organicconsumers
.org/campaigns/care-what-you-wear
Food & Fibers Project: www.foodand
fibersproject.com

University & Incubator Opportunities

California College of Arts (Professors Lynda
Grose, Amy Campos, and Sasha Duerr)
Fashion Institute of Design & Merchandising
(FIDM) (Janice Parades)
Fashion Institute of Technology (FIT)
(Jeffrey Silberman)
Parsons School of Design (Professors Timo
Rasmussen and Laura Parsons)

Fiber Festivals

These are events where you can meet farmers
and ranchers in person and in most cases
purchase their farm yarns directly from them.

NORTHEAST

Adirondack Wool & Arts Festival (NY)
CNY Fiber Arts Festival (NY)
Common Ground Country Fair (ME)
Connecticut Sheep, Wool & Fiber Festival (CT)
Fiber Fest 2017 (MD)
Fiber Festival of New England (MA)
Finger Lakes Fiber Arts Festival (NY)
Garden State Sheep Breeders Sheep and Fiber
Festival (NJ)
Maine Fiber Frolic (ME)
Maryland Alpaca and Fleece Festival (MD)
Maryland Sheep & Wool Festival (MD)
Massachusetts Sheep and Woolcraft Fair (MA)
New England Fiber Arts Summit (VT)
New Hampshire Sheep & Wool Festival (NH)
New York State Sheep & Wool Festival (NY)

Pennsylvania Endless Mountains Fiber
Festival (PA)
Rhode Island Fiber Festival and Craft Fair at
Coggeshall Farm Museum (RI)
Steel City Fiber Fest (PA)
Vermont Sheep and Wool Festival (VT)
Washington County Fiber Tour (NY)
Waynesburg Sheep & Fiber Festival (PA)
Western New York Fiber Arts Festival (NY)

MIDWEST

Autumn Fiber Festival (OH)
Fiber "U" (MO)
Fiber Arts Festival (ND)
Fiber Expo (MI)
Fiber in the Park (IL)
Fosston Fiber Festival (MN)
Great Lakes Fiber Show (OH)
Hoosier Hills Fiber Festival (IN)
Indiana Fiber & Music Festival (IN)
Iowa Sheep & Wool Festival (IA)
Jay County Fiber Arts Festival (IN)
Michigan Fiber Festival (MI)
Mid-Ohio Fiber Fair (OH)
Mid-Plains Fiber Fair (NE)
North Country Fiber Fair (SD)
Northern Michigan Lamb & Wool
Festival (MI)
Ozark Fiber Fling (MO)
Scotts Bluff Valley Fiber Fair (NE)
Shepherd's Harvest Festival (MN)
Shepherds Market Fiber Festival (IA)
Spring Fiber Fling (IL)
The Fiber Event (IN)
Tip of the Mitt Fiber Fair (MI)
Upper Valley Fiber Fest (OH)
Winter Woolfest (KS)
Wisconsin Alpaca & Fiber Fest (WI)
Wisconsin Sheep & Wool (WI)
Wool Gathering at Young's (OH)

SOUTH

Carolina Fiber Fest (NC)
DFW Fiber Fest (TX)
East Side Fiber Festival (TN)
Fall Fiber Festival and Montpelier Sheep Dog
Trials (VA)

Fiber Christmas in July (OK)
Fiber in the 'Boro (TN)
Florida Sheep, Wool, and Herding Dog
    Festival (FL)
Kentucky Sheep and Fiber Festival (KY)
Kentucky Wool Festival (KY)
Kid 'n Ewe (TX)
Middle Tennessee Fiber Festival (TN)
Olde Liberty Fibre Faire (VA)
Shenandoah Valley Fiber Festival (VA)
Southeastern Animal Fiber Fair (NC)

WEST

Alpacapalooza (WA)
Barn to Yarn (CA)
Black Sheep Gathering (OR)
California Wool & Fiber Festival (CA)
Casari Ranch Wool Festival (CA)

Elizabeth Fiber Festival (CO)
Estes Park Wool Market & Fiber Festival (CO)
Fiberfest Eureka!! (MT)
Flag Wool & Fiber Festival (AZ)
Lambtown (CA)
Mat Su Valley Fiber Festival (AK)
Okfiberfestival (WA)
Oregon Flock & Fiber Festival (OR)
Salida Fiber Festival (CO)
Sheep Is Life Celebration (AZ)
Shepherds' Extravaganza (WA)
Sneffles Fiber Festival (CO)
Sustainable Cotton Project, Farm Tours (CA)
The Black Sheep Gathering (OR)
The Taos Wool Festival (NM)
Trailing of the Sheep Festival (ID)
Vista Fiber Arts Fiesta (CA)
Wool Symposium (CA)

Suggested Reading

Adrosko, Rita. *Natural Dyes and Home Dyeing*. Mineola, NY: Dover, 1971.

Anderson, Sarah, and Judith Mackenzie. *The Spinner's Book of Yarn Designs: Techniques for Creating 80 Yarns*. North Adams, MA: Storey Publishing, 2013.

Beckert, Sven. *Empire of Cotton: A Global History*. New York: Random House, 2014.

Black, Kate. *Magnifeco: Your Head-to-Toe to Ethical Fashion and Non-Toxic Beauty*. Gabriola Island, BC: New Society, 2015.

Black, Sandy. *Eco-Chic: The Fashion Paradox*. London: Black Dog Publishing, 2011.

Black, Sandy. *The Sustainable Fashion Handbook*. London: Thames and Hudson, 2013.

Blackburn, Richard. *Sustainable Textiles: Life Cycle and Environmental Impact*. Oxford, UK: Woodhead Publishing, 2009.

Briggs, Raleigh. *Fix Your Clothes: The Sustainable Magic of Mending, Patching, and Darning*. Portland, OR: Microcosm Publishing, 2017.

Brown, Gabe. *Dirt to Soil: One Family's Journey into Regenerative Agriculture*. White River Junction, VT: Chelsea Green, 2018.

Brown, Sass. *Eco-Fashion*. London: Lawrence King Publishing, 2010.

Brown, Sass. *Refashioned: Cutting-Edge Clothing from Upcycled Materials*. London: Lawrence King Publishing, 2013.

Buchanan, Rita. *A Dyer's Garden: From Plant to Pot, Growing Dyes for Natural Fibers*. Loveland, CO: Interweave Press, 1995.

Burgess, Rebecca. *Harvesting Color: How to Find Plants and Make Natural Dyes*. New York: Artisan, 2011.

Church, George. *Regenesis: How Synthetic Biology Will Reinvent Nature and Ourselves*. New York: Basic Books, 2012.

Cline, Elizabeth. *Overdressed: The Shockingly High Cost of Cheap Fashion*. New York: Penguin, 2013.

Dean, Jenny. *Wild Color: The Complete Guide to Making and Using Natural Dyes*. New York: Potter Craft / Random House, 2010.

Desnos, Rebecca. *Botanical Colour at Your Fingertips*. London: Desnos, 2016.

Duerr, Sasha. *The Handbook of Natural Plant Dyes: Personalize Your Craft with Organic Colors from Acorns, Blackberries, Coffee, and Other Everyday Ingredients*. Portland, OR: Timber Press, 2011.

Duerr, Sasha. *Natural Color: Vibrant Plant Dye Projects for Your Home and Wardrobe*. New York: Watson-Guptill / Penguin, 2016.

Ekarius, Carol. *The Fleece & Fiber Sourcebook: More than 200 Fibers, from Animal to Spun Yarn*. North Adams, MA: Storey Publishing, 2011.

Ekarius, Carol. *The Field Guide to Fleece: 100 Sheep Breeds & How to Use Their Fibers*. North Adams, MA: Storey Publishing, 2013.

Fletcher, Kate. *Sustainable Fashion and Textiles: Design Journeys*. 2nd ed. New York: Routledge, 2014.

Fletcher, Kate. *Craft of Use: Post-Growth Fashion*. New York: Routledge, 2016.

Fletcher, Kate, and Lynda Grose. *Fashion and Sustainability: Design for Change*. London: Lawrence King Publishing, 2012.

Fletcher, Kate, and Mathilda Tham. *Routledge Handbook of Sustainability and Fashion*. New York: Routledge, 2015.

Flint, India. *Eco Colour: Botanical Dyes for Beautiful Textiles*. Loveland, CO: Interweave Press, 2010.

Flintoff, John-Paul. *Through the Eye of a Needle: The True Story of a Man Who Went*

Searching for Meaning and Ended Up Making His Y-Fronts. Hampshire, UK: Permanent Publications, 2010.

Fournier, Nola. *In Sheep's Clothing: A Hand-spinner's Guide to Wool.* Loveland, CO: Interweave Press, 1995.

Fox, Alice. *Natural Processes in Textile Art: From Rust-Dyeing to Found Objects.* London: Batsford/Pavilion, 2015.

Franquemont, Abby. *Respect the Spindle: Spin Infinite Yarns with One Amazing Tool.* Loveland, CO: Interweave Press, 2009.

Gardetti, Miguel Angel, and Ana Laura Torres. *Sustainability in Fashion and Textiles: Values, Design, Production and Consumption.* New York: Routledge, 2017.

Gordon, Jennifer Farley, and Colleen Hill. *Sustainable Fashion: Past, Present and Future.* London: Bloomsbury, 2015.

Gullingsrud, Annie. *Fashion Fibers: Designing for Sustainability.* London: Bloomsbury, 2017.

Lambert, Eva. *The Complete Guide to Natural Dyeing: Techniques and Recipes for Dyeing Fabrics, Yarns, and Fibers at Home.* Loveland, CO: Interweave Press, 2010.

McLaughlin, Chris. *A Garden to Dye For: How to Use Plants from the Garden to Create Natural Colors for Fabrics & Fibers.* Pittsburgh: St. Lynn's Press, 2014.

Meuret, Michel, and Fred Provenza. *The Art and Science of Shepherding: Tapping the Wisdom of French Herders.* Austin, TX: Acres USA, 2014.

Minney, Safia. *Slave to Fashion.* Northampton, UK: New Internationalist, 2017.

Pollan, Michael. *The Botany of Desire: A Plant's-Eye View of the World.* New York: Random House, 2001.

Rivoli, Pietra. *Travels of a T-Shirt in the Global Economy: An Economist Examines the Markets, Power, and Politics of World Trade.* Hoboken, NJ: Wiley and Sons, 2009.

Rodabaugh, Katrina. *Mending Matters: Stitch, Patch, and Repair Your Favorite Denim & More.* New York: Abrams Books, 2018.

Savory, Allan, and Jody Butterfield. *Holistic Management: A Commonsense Revolution to Restore Our Environment.* Washington, DC: Island Press, 2016.

Schor, Juliet. *True Wealth: How and Why Millions of Americans Are Creating a Time-Rich, Ecologically Light, Small-Scale, High-Satisfaction Economy.* New York: Penguin, 2011.

Smith, Beth. *The Spinner's Book of Fleece: A Breed-by-Breed Guide to Choosing and Spinning the Perfect Fiber for Every Purpose.* North Adams, MA: Storey Publishing, 2014.

Steingraber, Sandra. *Living Downstream: An Ecologist's Personal Investigation of Cancer and the Environment.* Boston: Da Capo Press, 2010.

Toensmeier, Eric. *The Carbon Farming Solution: A Global Toolkit of Perennial Crops and Regenerative Agriculture Practices for Climate Change Mitigation and Food Security.* White River Junction, VT: Chelsea Green, 2016.

Vejar, Kristine. *The Modern Natural Dyer: A Comprehensive Guide to Dyeing Silk, Wool, Linen, and Cotton at Home.* New York: Abrams, 2015.

Voisin, André. *Grass Productivity.* Washington, DC: Island Press, 1989.

Wellesley-Smith, Claire. *Slow Stitch: Mindful and Contemplative Textile Art.* London: Batsford/Pavilion, 2015.

Wilkes, Stephany. *Raw Material: Working Wool in the West.* Corvallis, OR: Oregon State University Press, 2018.

Notes

Introduction

1. Barbara A. Cohn, Piera M. Cirillo, and Mary Beth Terry, "DDT and Breast Cancer: Prospective Study of Induction Time and Susceptibility Windows," *Journal of the National Cancer Institute* djy198 (February 2019), https://doi.org/10.1093/jnci/djy198.

Chapter One: The Cost of Our Clothes

1. Ellen MacArthur Foundation, *A New Textiles Economy: Redesigning Fashion's Future* (EMF, 2017), 18, http://www.ellen macarthurfoundation.org.

2. Ellen MacArthur Foundation, *A New Textiles Economy*, 38.

3. Rita Kant, "Textile Dyeing Industry an Environmental Hazard," *Natural Science* 4, no. 1 (2012): 22–26, https://doi.org /10.4236/ns.2012.41004.

4. European Parliament, *Workers' Conditions in the Textile and Clothing Sector: Just an Asian Affair? Issues at Stake After the Rana Plaza Tragedy* (Brussels: EU, 2014), 1–9, http://www.europarl.europa.eu/EPRS /140841REV1-Workers-conditions-in-the -textile-and-clothing-sector-just-an-Asian -affair-FINAL.pdf.

5. Asia Floor Wage Alliance, *Precarious Work in the H&M Global Value Chain* (AFWA, 2016), 34–35, https://asia.floorwage.org /workersvoices/reports/precarious-work -in-the-h-m-global-value-chain/view.

6. International Labour Office, *Marking Progress Against Child Labour: Global Estimates and Trends 2000–2012* (Geneva: ILO, 2013), vii, http://www.ilo.org/wcmsp5 /groups/public/---ed_norm/---ipec /documents/publication/wcms_221513.pdf.

7. Davuluri Venkateswarlu, *Cotton's Forgotten Children: Child Labour and Below Minimum Wages in Hybrid Cottonseed Production in India* (Utrecht: India Committee of the Netherlands, 2015), 7, http://www.indianet.nl/pdf/Cottons ForgottenChildren.pdf.

8. Elizabeth Cline, *Overdressed: The Shockingly High Cost of Cheap Fashion* (New York: Penguin Press, 2012).

9. Greenpeace, *Timeout for Fast Fashion* (Hamburg: Greenpeace, 2015), 2, https:// www.greenpeace.org/archive-international /Global/international/briefings/toxics /2016/Fact-Sheet-Timeout-for-fast -fashion.pdf.

10. Lydia McAllister, "Textile Waste by the Numbers," *Vox*, March 24, 2016, https:// www.voxmagazine.com/news/textile-waste -by-the-numbers/article_9ea228ba-f13a- 11e5-8c76-5b50180f85de.html.

11. Ellen MacArthur Foundation, *A New Textiles Economy*, 73.

12. Greenpeace, *Timeout for Fast Fashion*, 6.

13. European Commission, *Environmental Improvement Potential of Textiles* (Seville: JRC Scientific and Technical Reports, 2006), 76, http://susproc.jrc.ec.europa.eu /textiles/docs/120423%20IMPRO% 20Textiles_Publication%20draft%20v1.pdf.

14. Bren Microplastics Project, "Microfiber Pollution and the Apparel Industry," Bren Microplastics (website), https://brenmicro plastics.weebly.com/project-findings.

15. Chelsea Rochman et al., "Anthropogenic Debris in Seafood: Plastic Debris and Fibers from Textiles in Fish and Bivalves Sold for Human Consumption," *Scientific*

Reports 5, no. 14340 (2015): 7, https://www. https://doi.org/10.1038/srep14340.

16. Lisbeth Cauwenberghe and Colin Janssen, "Microplastics in Bivalves Cultured for Human Consumption," *Environmental Pollution* 193 (2014): 65, http://dx.doi.org/10.1016/j.envpol.2014.06.010.

17. M. A. Toups et al., "Origin of Clothing Lice Indicates Early Clothing Use by Anatomically Modern Humans in Africa," *Molecular Biology and Evolution* 28, no. 1 (2011): 29–32, https://doi.org/10.1093/molbev/msq234.

18. Sandra Steingraber, "Sandra Steingraber's War on Toxic Trespassers," interviewed by Bill Moyers, April 19, 2013, video, http://billmoyers.com/segment/sandra-steingrabers-war-on-toxic-trespassers/.

19. Sidley Environmental Practice, "Significant Changes to TSCA Will Affect a Broad Range of Companies That Manufacture and Use Chemicals," *Sidley Update*, June 9, 2016, https://www.sidley.com/-/media/update-pdfs/2016/06/20160608-environmental-update.pdf.

20. Steingraber, "Sandra Steingraber's War on Toxic Trespassers."

21. A. Pruss-Ustan et al., *Preventing Disease Through Healthy Environments: A Global Assessment of the Burden of Disease from Environmental Risks* (Geneva: World Health Organization, 2016), viii, http://apps.who.int/iris/bitstream/10665/204585/1/9789241565196_eng.pdf?ua=1.

22. United Nations, *Report of the United Nations Conference on Sustainable Development: Rio + 20, 20–22 June 2012* (New York: United Nations, 2012), 27, https://rio20.un.org/sites/rio20.un.org/files/a-conf.216l-1_english.pdf.pdf.

23. Pruss-Ustan, *Preventing Disease Through Healthy Environments*, xxiii.

24. Steingraber, "Sandra Steingraber's War on Toxic Trespassers."

25. Lake Research Partners, *Key Survey Findings on Laws Governing Toxic Chemicals* (Los Angeles: LRP, 2009), 2, http://saferchemicals.org/sc/wp-content/uploads/pdf/Lake_Research_Findings.pdf.

26. Heinrich Zollinger, *Color Chemistry: Syntheses, Properties, and Applications of Organic Dyes and Pigments* (New York: VCH Publishers, 1987), 92–102.

27. Farah Maria Drumond Chequer et al., "Textile Dyes: Dyeing Process and Environmental Impact," in *Eco-Friendly Textile Dyeing and Finishing*, ed. Melih Gunay (London: InTech Open, 2013), 152, https://doi.org/10.5772/53659; K. T. Chung and C. E. Cerniglia, "Mutagenicity of Azo Dyes: Structure-Activity Relationships," *Mutation Research* 277, no. 3 (1992): 201–20, https://doi.org/10.1016/0165-1110(92)90044-A.

28. Greenpeace, *Chemistry for Any Weather: Greenpeace Tests Outdoor Clothes for Perfluorinated Toxins* (Hamburg: Greenpeace, 2012), 43, https://www.greenpeace.org/romania/Global/romania/detox/Chemistry%20for%20any%20weather.pdf.

29. C. J Ogugbue and T. Sawidis, "Bioremediation and Detoxification of Synthetic Wastewater Containing Triarylmethane Dyes by Aeromonas Hydrophila Isolated from Industrial Effluent," *Biotechnology Research International*, article 967925 (2011): 1, http://dx.doi.org/10.4061/2011/967925.

30. Martin A. Hubbe et al., "Cellulosic Substrates for Removal of Pollutants from Aqueous Systems: A Review. 2. Dyes," *BioResources* 7, no. 2 (2012): 2593, http://ojs.cnr.ncsu.edu/index.php/BioRes/article/view/BioRes_07_2_2592_Hubbe_BO_Cellulosic_Substrates_Remov_Pollutants_Review_Pt2_Dyes/1530.

31. P. A. Carneiro et al., "Assessment of Water Contamination Caused by a Mutagenic Textile Effluent / Dyehouse Effluent Bearing Disperse Dyes," *Journal of Hazardous Materials* 174, nos. 1–3 (2010): 696, https://doi.org/10.1016/j.jhazmat.2009.09.106.

32. Personal communication.

33. Birgitta Kütting et al., "Monoarylamines in the General Population—A Cross-Sectional Population-Based Study Including 1004 Bavarian Subjects," *International Journal of Hygiene and Environmental Health* 212, no. 3 (May 2009): 301, https://doi.org /10.1016/j.ijheh.2008.07.004.

34. Hendrick Modick et al., "Ubiquitous Presence of Paracematol in Human Urine: Sources and Implications," *Reproduction* 147 (2014): 105, https://doi.org/10.1530 /REP-13-0527.

35. Shanna Swan et al., "Decrease in Anogenital Distance Among Male Infants with Prenatal Phthalate Exposure," *Environmental Health Perspectives* 113, no. 8 (2005): 1056–61, https://doi.org/10.1289/ehp.8100; Shanna Swan et al., "Prenatal Phthalate Exposure and Reduced Masculine Play in Boys," *International Journal of Andrology* 33, no. 2 (2010): 259–69, https://dx.doi .org/10.1111%2Fj.1365-2605.2009.01019.x; Richard Grady and Shella Sathyanarayana, "An Update on Phthalates and Male Reproductive Development and Function," *Current Urology Reports* 13, no. 4 (2012): 307–10, https://doi.org/10.1007/s11934 -012-0261-1.

36. Personal communication.

37. E. Diamanti-Kandarakis et al., "Endocrine-Disrupting Chemicals: An Endrocrine Society Scientific Statement," *Endocrine Reviews* 30, no. 4 (2009): 293–342, https://doi.org/10.1210/ er.2009-0002; European Commission, *European Union Risk Assessment Report—Bis (2-ethylhexyl) Phthalate (DEHP)* (Ispra, Italy: European Chemicals Bureau, 2008), https://ec.europa.eu/jrc /en/publication/eur-scientific-and -technical-research-reports/european -union-risk-assessment-report-bis-2 -ethylhexyl-phthalate-dehp; and European Environment Agency, "The Impacts of Endocrine Disruptors on Wildlife, People and Their Environments," EEA Technical Report no. 2 (2012), http://www.eea .europa.eu/publications/the-impacts-of -endocrine-disrupters.

38. World Health Organization, *State of the Science of Endocrine Disrupting Chemicals—2012* (Geneva: WHO/UNEP, 2012), http://www.who.int/ceh/publications /endocrine/en/.

39. Adrian Burton, "Study Suggests Long-Term Decline in French Sperm Quality," *Environmental Health Perspectives* 121, no. 2 (2013): A46, https://doi.org /10.1289/ehp.121-a46.

40. European Environment Agency, "The Impacts of Endocrine Disruptors," 8.

41. European Environment Agency, "The Impacts of Endocrine Disruptors," 92.

42. World Health Organization, *State of the Science of Endocrine Disrupting Chemicals*, xiv.

43. World Health Organization, *State of the Science of Endocrine Disrupting Chemicals*, 14.

44. Juliet B. Schor, *True Wealth: How and Why Millions of Americans Are Creating a Time-Rich, Ecologically Light, Small-Scale, High-Satisfaction Economy* (New York: Penguin, 2011).

45. Greenpeace, *Timeout for Fast Fashion*, 5.

46. Elizabeth Cline, "In Trendy World of Fast Fashion Styles Aren't Made to Last," interview by Jim Zarroli, *All Things Considered*, NPR, March 11, 2013, audio, https://www.npr.org/2013/03/11 /174013774/in-trendy-world-of-fast -fashion-styles-arent-made-to-last.

47. Ellen MacArthur Foundation, *A New Textiles Economy*, 21.

48. Cline, "In Trendy World of Fast Fashion Styles Aren't Made to Last," interview by Jim Zarroli.

49. Human Rights Watch, *Whoever Raises Their Head Suffers the Most: Workers' Rights in Bangladesh's Garment Factories* (Amsterdam: HRW, 2015), 6, https:// www.hrw.org/report/2015/04/22/whoever -raises-their-head-suffers-most/workers -rights-bangladeshs-garment.

50. Greenpeace, *Dirty Laundry: Unraveling the Corporate Connections to Toxic Water Pollution in China* (Hamburg: Greenpeace, 2015), 5, https://www.greenpeace .org/international/publication/7168 /dirty-laundry/.

51. Greenpeace, *Destination Zero: Seven Years of Detoxing the Fashion Industry* (Hamburg: Greenpeace, 2018), 8, https://www .greenpeace.org/international/publication /17612/destination-zero/.

52. Katrina Rodabaugh, "Make Thrift Mend" (website), https://www.katrinarodabaugh .com/make-thrift-mend.

53. Kate Fletcher, *Craft of Use: Post-Growth Fashion* (London: Routledge, 2016).

Chapter Three: Soil-to-Soil Clothing and the Carbon Cycle

1. Eric Toensmeier, *The Carbon Farming Solution: A Global Toolkit of Perennial Crops and Regenerative Agriculture Practice for Climate Change Mitigation and Food Security* (White River Junction, VT: Chelsea Green, 2016).

2. Doug Gurian-Sherman, *Raising the Steaks: Global Warming and Pasture-Raised Beef Production in the United States* (Cambridge, MA: Union of Concerned Scientists, 2011), https://www.ucsusa.org /sites/default/files/legacy/assets/documents /food_and_agriculture/global-warming -and-beef-production-report.pdf.

3. Kees Jan van Groenigen et al., "Faster Decomposition Under Increased Atmospheric CO_2 Limits Soil Carbon Storage," *Science* 244 (May 2014): 508–09, https:// doi.org/10.1126/science.1249534.

4. Rebecca Ryals and Whendee Silver, "Effects of Organic Matter Amendments on Net Primary Production and Greenhouse Gas Emissions in Annual Grasslands," *Ecological Applications* 23, no. 1 (2013): 56, https://doi.org/10.1890/12-0620.1.

5. Personal communication.

6. L. E. Flint et al., *Increasing Soil Organic Carbon to Mitigate Greenhouse Gases and Increase Climate Resiliency for California: A Report for California's Fourth Climate Change Assessment* (Sacramento: California Department of Natural Resources, 2018), http://www.climateassessment .ca.gov/techreports/docs/20180827 -Agriculture_CCCA4-CNRA-2018-006.pdf.

7. Richard Teague et al., "The Role of Ruminants in Reducing Agriculture's Carbon Footprint in North America," *Journal of Soil and Water Conservation* 71, no. 2 (2016): 160, https://doi.org/10.2489 /jswc.71.2.156.

8. Ronald Follett and Debbie Reed, "Soil Carbon Sequestration in Grazing Lands: Societal Benefits and Policy Implications," *Range Ecology and Management* 63, no. 1 (2010): 4–15, https://doi.org/10.2111/08-225.1.

9. Douglas Frank, Samuel McNaughton, and Benjamin Tracy, "The Ecology of the Earth's Grazing Ecosystems," *BioScience* 48, no. 7 (1998): 513–28, https://doi.org / 10.2307/1313313.

10. Richard Teague et al., "Multi-Paddock Grazing on Rangelands: Why the Perceptual Dichotomy Between Research Results and Rancher Experience?" *Journal of Environmental Management* 128 (2013): 699–717, https://doi.org/10.1016/j .jenvman.2013.05.064.

11. Teague et al., "The Role of Ruminants in Reducing Agriculture's Carbon Footprint in North America," 156.

12. Megan Machmuller et al., "Emerging Land Use Practices Rapidly Increase Soil Organic Matter," *Nature Communications* 6, article 6995 (2015), https://doi.org/10.1038 /ncomms7995; D. G. Milchunas and W. K. Lauenroth, "Quantitative Effects of Grazing on Vegetation and Soils Over a Global Range of Environments," *Ecological Monographs* 63, no. 4 (1993): 327–66, https://doi.org/10.2307/2937150; S. Vetter et al., "Effects of Land Tenure, Geology and Topography on Vegetation and Soils of Two Grassland Types in South Africa," *African Journal of Range and Forage Science* 23,

no. 1 (2006): 13–27, https://doi.org /10.2989/10220110609485883.

13. *The Art and Science of Shepherding: Tapping the Wisdom of French Herders*, edited by Michel Meuret and Fred Provenza (Austin, TX: Acres Press, 2014).

14. Christina Figueres, *Sequestering Carbon in Soil Addressing the Climate Threat*, prepared remarks, Paris, May 2017, https://agrinatura-eu.eu/2017/05 /sequestering-carbon-in-soil-addressing -the-climate-threat.

15. Personal communication.

16. Jeffrey Creque, *Bare Ranch Carbon Farm Plan* (Marin, CA: Carbon Cycle Institute, 2016).

17. Rebecca Ryals et al., "Impacts of Organic Matter Amendments on Carbon and Nitrogen Dynamics in Grassland Soils," *Soil Biology and Biochemistry* 78 (2014): 52–61, https://doi.org/10.1016/j.soilbio .2013.09.011.

18. Marcia DeLonge et al., "A Lifecycle Model to Evaluate Carbon Sequestration Poten- tial and Greenhouse Gas Dynamics of Managed Grassland," *Ecosystems* 16 (2013): 962–79, https://doi.org/10.1007 /s10021-013-9660-5.

Chapter Four: The False Solution of Synthetic Biology

1. ETC Group, "Synthetic Biology," ETC Group (website), https://www.etcgroup.org /issues/synthetic-biology.

2. Eduardo Barretto de Figueiredo et al., "Greenhouse Gas Emission Associated with Sugar Production in Southern Brazil," *Carbon Balance and Management* 5, no. 3 (June 2010): 1, https://doi.org/10.1186 /1750-0680-5-3.

3. World Wildlife Fund, "Sustainable Agriculture: Sugarcane," WWF (website), https://www.worldwildlife.org/industries /sugarcane.

4. Institute for Responsible Technology, "FAQS," IRT (website), https://responsible technology.org/gmo-education/faqs.

5. Janet Cotter and Dana Perls, *Gene-Edited Organisms in Agriculture: Risks and Unexpected Consequences* (Amsterdam: Friends of the Earth, 2018), 4, http://foe .org/wp-content/uploads/2018/09/FOE _GenomeEditingAgReport_final.pdf.

6. Sharon Begley, "CRISPR-Edited Cells Linked to Cancer Risk in 2 Studies," *Scientific American*, STAT Biotech, June 12, 2018, shttps://www.scientificamerican .com/article/crispr-edited-cells-linked-to -cancer-risk-in-2-studies/.

7. Sharon Begley, "Potential DNA Damage from CRISPR 'Seriously Underestimated,' Study Finds," *Scientific American*, STAT Medical and Biotech, July 16, 2018, https://www.scientificamerican.com /article/potential-dna-damage-from-crispr -seriously-underestimated-study-finds.

8. Vandana Shiva, "Vandana Shiva on Seed Monopolies, GMOs, and Farmer Suicides in India," GM Watch (website), 2013, http://gmwatch.org/en/news/archive /2013/15165-vandana-shiva-on-seed -monopolies-gmos-and-farmer-suicides -in-india.

9. Personal communication.

10. Celine Gasnier et al., "Glyphosate-Based Herbicides Are Toxic and Endocrine Disruptors in Human Cell Lines," *Toxicol- ogy* 262 (2009): 184, https://doi.org/10 .1016/j.tox.2009.06.006.

11. Cristina Casals-Casas and Béatrice Des- vergne, "Endocrine Disruptors: From Endocrine to Metabolic Disruption," *Annual Review of Physiology* 73, no. 1 (2011): 135–62, https://doi.org/10.1146 /annurev-physiol-012110-142200.

12. Elizabeth A. Scribner et al., "Concentra- tions of Glyphosate, Its Degradation Product, Aminomethylphophonic Acid, and Glufosinate in Ground- and Sur- face-Water, Rainfall, and Soil Samples Collected in the United States, 2001–06," US Geological Survey Scientific Investiga- tions Report 5122 (2007), https://pubs .usgs.gov/sir/2007/5122.

13. David Quist and Ignacio Chapela, "Transgenic DNA Introgressed into Traditional Maize Landraces in Oaxaca, Mexico," *Nature* 414 (November 2001): 541, https://doi.org/10.1038/35107068.

14. A. Pineyro-Nelson et al., "Transgenes in Mexican Maize: Molecular Evidence and Methodological Considerations for GMO Detection in Landrace Populations," *Molecular Ecology* 18, no. 4 (2009): 755, https://doi.org/10.1111/j.1365-294X.2008.03993.x.

15. Emily Hazzard, "A Conversation with Wes Jackson, President of the Land Institute," *Atlantic*, March 23, 2011, https://www.theatlantic.com/international/archive/2011/03/a-conversation-with-wes-jackson-president-of-the-land-institute/72927.

16. George Church and Ed Regis, *Regenesis: How Synthetic Biology Will Reinvent Nature and Ourselves* (New York: Basic Books, 2012).

17. European Court of Justice, "Organisms Obtained by Mutagensis Are GMOs and Are, in Principle, Subject to the Obligations Laid Down by the GMO Directive," ECJ Press Release no. 111/18 (2018), https://curia.europa.eu/jcms/upload/docs/application/pdf/2018-07/cp180111en.pdf.

18. Diana Crow, "Six Amazing Things to Watch for in Synthetic Biology," *Neo.Life* (blog), October 12, 2017, https://medium.com/neodotlife/6-things-to-watch-in-synthetic-biology-f76666c7114e.

19. Gabe Brown, *Dirt to Soil: One Family's Journey into Regenerative Agriculture* (White River Junction, VT: Chelsea Green, 2018).

20. Gregory Mandel and Gary Marchant, "The Living Regulatory Challenges of Synthetic Biology," *Iowa Law Review* 100, rev. 155 (2014), https://ilr.law.uiowa.edu/print/volume-100-issue-1/the-living-regulatory-challenges-of-synthetic-biology/.

21. Cotter and Perls, *Gene-Edited Organisms in Agriculture*, 5.

22. Michael Eisenstein, "Living Factories of the Future," *Nature* 531 (March 2016): 403, https://doi.org/10.1038/531401a.

23. The White House, "National Bioeconomy Blueprint" (Washington, DC: White House, 2012), 1, https://obamawhitehouse.archives.gov/sites/default/files/microsites/ostp/national_bioeconomy_blueprint_april_2012.pdf.

24. ETC Group and Fibershed, *Genetically Engineered Clothes: Synthetic Biology's New Spin on Fast Fashion* (ETC and Fibershed, 2018), 13, https://www.etcgroup.org/sites/www.etcgroup.org/files/files/genetically_engineered_clothes_etc_fibershed_web.pdf.

25. Presidential Commission for the Study of Bioethical Issues, *New Directions: The Ethics of Synthetic Biology and Emerging Technologies* (2010), 62, https://bioethicsarchive.georgetown.edu/pcsbi/sites/default/files/PCSBI-Synthetic-Biology-Report-12.16.10_0.pdf.

26. Antonio Regalado, "CRISPR Inventor Feng Zhang Calls for a Moratorium on Gene-Edited Babies," *MIT Technology Review*, November 26, 2018, https://www.technologyreview.com/s/612465/crispr-inventor-feng-zhang-calls-for-moratorium-on-baby-making/.

27. Jon Cohen, "I Feel an Obligation to Be Balanced: Noted Biologist Comes to the Defense of Gene Editing Babies," *Science*, November 28, 2018, https://www.sciencemag.org/news/2018/11/i-feel-obligation-be-balanced-noted-biologist-comes-defense-gene-editing-babies.

Chapter Five: Implementing the Vision with Plant-Based Fibers

1. Congressional Research Service, "California Agricultural Production and Irrigated Water Use," CRS 7-5700 (June 30, 2015), summary, https://fas.org/sgp/crs/misc/R44093.pdf.

2. Sven Beckert, *Empire of Cotton: A Global History* (New York: Vintage Books, 2014), xv.

3. EcoChoices, "Conventional Cotton Statistics," EcoChoices (website), http://www.ecochoices.com/1/cotton_statistics.html.

4. The Non-GMO Project, "Cotton," Non-GMO Project (website), https://www.nongmoproject.org/high-risk/cotton.

5. Dan Charles, "Not Just for Cows Anymore: New Cottonseed Is Safe for People to Eat," *All Things Considered*, NPR, October 17, 2018, https://www.npr.org/sections/thesalt/2018/10/17/658221327/not-just-for-cows-anymore-new-cottonseed-is-safe-for-people-to-eat.

6. Know The Chain, *Apparel and Footwear Benchmark Findings Report* (San Francisco: KTC, 2016), 6, https://knowthechain.org/wp-content/plugins/ktc-benchmark/app/public/images/benchmark_reports/KTC_A&F_ExternalReport_Final.pdf.

7. Anti-Slavery, "Cotton Crimes: Uzbekistan and Turkmenistan," Anti-Slavery (website), https://www.antislavery.org/what-we-do/uzbekistan-turkmenistan.

8. As You Sow, "Responsible Sourcing Network: A Project of As You Sow," Responsible Sourcing Network (website), https://www.sourcingnetwork.org.

9. R. Zetterstrom, "Child Health and Environmental Pollution in the Aral Sea Region of Kazakhstan," *Acta Paediatrica* 88, no. s429 (1999): 49–54, https://doi.org/10.1111/j.1651-2227.1999.tb01290.x.

Chapter Six: Implementing the Vision with Animal Fibers and Mills

1. Personal communication.

2. Jeff Price, "2016 State of US Textile Industry," Textile World (website), http://www.textileworld.com/textile-world/features/2016/05/2016-state-of-the-u-s-textile-industry.

3. Chris Kennedy, "Human Consumption of Earth's Natural Resources Has Tripled in 40 Years," Sierra Club Grassroots Network (website), August 13, 2016, https://content.sierraclub.org/grassrootsnetwork/team-news/2016/08/human-consumption-earths-natural-resources-has-tripled-40-years.

4. FAO, "The State of Food and Agriculture in 2014: Innovation in Family Farming," *FAO In Brief* (2014), 2, http://www.fao.org/3/a-i4036e.pdf.

Conclusion: A Future Based in Truth

1. Edward E. Baptist, *The Half Has Never Been Told: Slavery and the Making of American Capitalism* (New York: Basic Books, 2014).

Appendix A: Ingredients to Watch

1. Sofie Christiansen et al., "Synergistic Disruption of External Male Sex Organ Development by a Mixture of Four Anti-Androgens," *Environmental Health Perspectives* 117, no. 12 (2009): 1839–46, https://dx.doi.org/10.1289%2Fehp.0900689.

2. Gasnier et al., "Glyphosate-Based Herbicides Are Toxic and Endocrine Disruptors."

3. Greenpeace, *A Red Card for Sportswear Brands* (Hamburg: Greenpeace, 2014), http://www.greenpeace.org/international/Global/international/publications/toxics/2014/Detox-Football-Report.pdf.

4. Ruthann Rudel et al., "Phthalates, Alkylphenols, Pesticides, Polybrominated Diphenyl Ethers, and Other Endocrine-Disrupting Compounds in Indoor Air and Dust," *Environmental Science and Technology* 37, no. 20 (2003): 4543–53, https://doi.org/10.1021/es0264596.

5. Centers for Disease Control and Prevention, *Third National Report on Human Exposure to Environmental Chemicals* (Washington, DC: US Department of Health and Human Services, Centers for Disease Control and Prevention, 2005), https://clu-in.org/download/contaminantfocus/pcb/third-report.pdf.

6. Shanna Swan, "Environmental Phthalate Exposure in Relation to Reproductive Outcomes and Other Health Endpoints in Humans," *Environmental Research* 108, no. 2 (2008): 177–84, https://www.ncbi

.nlm.nih.gov/pubmed/18949837; Katharina M. Main et al., "Human Breast Milk Contamination with Phthalates and Alterations of Endogenous Reproductive Hormones in Infants Three Months of Age," *Environmental Health Perspectives* 114, no. 2 (2006): 270–76, https://dx.doi.org/10.1289%2Fehp.8075; Po-chin Huang et al., "Association Between Prenatal Exposure to Phthalates and the Health of Newborns," *Environment International* 35, no. 1 (2008): 14–20, https://doi.org/10.1016/j.envint.2008.05.012.

7. Grady and Sathyanarayana, "An Update on Phthalates and Male Reproductive Development and Function."; C. Wohlfahrt-Veje, K. M. Main, and N. E. Skakkebaek, "Testicular Dysgenesis Syndrome: Foetal Origin of Adult Reproductive Problems," *Clinical Endocrinology* 71, no. 4 (2009): 459–65, https://doi.org/10.1111/j.1365-2265.2009.03545.x.

8. Susan M. Duty et al., "Phthalate Exposure and Human Semen Parameters," *Epidemiology* 14, no. 3 (2003): 269–77, https://doi.org/10.1097/01.EDE.0000059950.11836.16; Giuseppe Latini et al., "Phthalate Exposure and Male Infertility," *Toxicology* 226, nos. 2–3 (2006): 90–98, https://doi.org/10.1016/j.tox.2006.07.011.

9. Tobias Frömel and Thomas P. Knepper, "Biodegradation of Fluorinated Alkyl Substances," *Reviews of Environmental Contamination and Toxicology* 208 (2010): 161–77, https://doi.org/10.1007/978-1-4419-6880-7_3.

10. Environmental Working Group, *Pollution in People: Cord Blood Contaminants in Minority Newborns* (Washington, DC: EWG, 2009), http://www.marylandpirg.org/sites/pirg/files/resources/2009-Minority-Cord-Blood-Report.pdf.

11. Greenpeace, *A Red Card for Sportswear Brands.*

12. Hermann Fromme et al., "Pre- and Postnatal Exposure to Perfluorinated Compounds (PFCs)," *Environmental Science and Technology* 44, no. 18 (2010): 7123–29, https://doi.org/10.1021/es101184f.

13. Julie B. Herbstman et al., "Birth Delivery Mode Modifies the Associations Between Prenatal Polychlorinated Biphenyl (PCB) and Polybrominated Diphenyl Ether (PBDE) and Neo-Natal Thyroid Hormone Levels," *Environmental Health Perspectives* 116, no. 10 (2008): 1376–82, https://dx.doi.org/10.1289%2Fehp.11379.

14. Benjamin Apelberg et al., "Determinants of Fetal Exposure to Polyfluoro-Alkyl Compounds in Baltimore, Maryland," *Environmental Science and Technology* 41, no. 11 (2007): 3891–97, https://doi.org/10.1021/es0700911.

15. Chunyaun Fei et al., "Maternal Levels of Perfluorinated Chemicals and Subfecundity," *Human Reproduction* 24, no. 5 (2009): 1200–05, https://doi.org/10.1093/humrep/den490; Kyle Steenland et al., "Association of Perfluorooctanoic Acid and Perfluorooctane Sulfonate with Serum Lipids Among Adults Living Near a Chemical Plant," *American Journal of Epidemiology* 170, no. 10 (2009), 1268–78, https://doi.org/10.1093/aje/kwp279; Ulla N. Joensen et al., "Do Perfluoroalkyl Compounds Impair Human Semen Quality?" *Environmental Health Perspectives* 117, no. 6 (2009): 923–27, https://dx.doi.org/10.1289%2Fehp.0800517.

16. Environmental Working Group, *Pollution in People.*

17. Robin Vestergren et al., "Dietary Exposure to Perfluoroalkyl Acids for the Swedish Population in 1999, 2005, and 2010," *Environment International* 49 (2012): 120–27, https://doi.org/10.1016/j.envint.2012.08.016.

18. SEPA, Scottish Environment Protection Agency, Accessed September 22, 2011, http://apps.sepa.org.uk/spripa/Pages/SubstanceInformation.aspx?pid=154

19. US Environmental Protection Agency, *Nonylphenol (NP) and Nonylphenol*

Ethoxylates (NPEs) Action Plan (Washington, DC: EPA, 2010), https://www.epa.gov/sites/production/files/2015-09/documents/rin2070-za09_np-npes_action_plan_final_2010-08-09.pdf.

20. Greenpeace, *A Red Card for Sportswear Brands*.

21. Yixiu Wu, "You Did It! Toxic Chemical Banned in EU Textile Imports," Greenpeace (website), July 22, 2015, http://www.greenpeace.org/international/en/news/Blogs/makingwaves/NPE-toxic-chemical-banned-EU-textile/blog/53582/.

22. US Environmental Protection Agency, *Nonylphenol (NP) and Nonylphenol Ethoxylates (NPEs)*.

23. US Environmental Protection Agency, *Ambient Aquatic Life Water Quality Criteria for Nonylphenol—Final* (Washington, DC: EPA Office of Water, 2005), https://nepis.epa.gov/Exe/ZyPDF.cgi/P1004WZW.PDF?Dockey=P1004WZW.PDF

24. Erin Ihde and Ji Meng Loh, "Association of Environmental Chemicals and Estrogen Metabolites in Children," *BMC Endocrine Disorders* 15, no. 83 (2015), https://doi.org/10.1186/s12902-015-0079-1.

25. David Case, "The Real Story Behind Bisphenol A," *Fast Company*, February 1, 2009, https://www.fastcompany.com/1139298/real-story-behind-bisphenol.

26. National Institute of Environmental Health Sciences, "NTP Finalizes Report on Bisphenol A," NIH Press Release, September 3, 2008, https://www.niehs.nih.gov/news/newsroom/releases/2008/september03/index.cfm.

27. Environmental Working Group, *Pollution in People*.

28. D. Li et al., "Occupational Exposure to Bisphenol-A (BPA) and the Risk of Self-Reported Male Sexual Dysfunction," *Human Reproduction* 100 (2009): 1–9, https://doi.org/10.1093/humrep/dep381.

29. American Chemical Society, "Hard Plastics Decomposing in Oceans, Releasing Endocrine Disrupter BPA," ACS Press Release, March 23, 2010, https://www.acs.org/content/acs/en/pressroom/newsreleases/2010/march/hard-plastics-decompose-in-oceans-releasing-erine-disruptor-bpa.html.

30. Heinrich Zollinger, *Color Chemistry: Syntheses, Properties, and Applications of Organic Dyes and Pigments* (New York: VCH Publishers, 1987), 92–102.

31. Chung and Cerniglia, "Mutagenicity of Azo Dyes."

32. Ogugbue and Sawidis, "Bioremediation and Detoxification of Synthetic Wastewater Containing Triarylmethane Dyes."

33. Hubbe et al., "Cellulosic Substrates for Removal of Pollutants from Aqueous Systems."

34. Carneiro et al., "Assessment of Water Contamination Caused by a Mutagenic Textile Effluent / Dyehouse Effluent Bearing Disperse Dyes."

35. World Health Organization, *State of the Science of Endocrine Disrupting Chemicals*.

36. Kütting et al., "Monoarylamines in the General Population—a Cross-Sectional Population-Based Study."

37. Swan et al., "Decrease in Anogenital Distance Among Male Infants with Prenatal Phthalate Exposure"; Swan et al., "Prenatal Phthalate Exposure and Reduced Masculine Play in Boys."

38. Charles Benbrook, "Trends in Glysophate Herbicide Use in the United States and Globally," *Environmental Sciences Europe* 28, no. 3 (2016), https://doi.org/10.1186/s12302-016-0070-0.

39. Gasnier et al., "Glyphosate-Based Herbicides Are Toxic and Endocrine Disruptors."

40. GreenMedInfo, "Glyphosate Induces Human Breast Cancer Cells Growth Via Estrogen Receptors," GreenMedInfo web site, June 7, 2013, http://www.greenmedinfo.com/article/glyphosate-induces-human-breast-cancer-cells-growth-estrogen-receptors.

41. Casals-Casas and Desvergne, "Endocrine Disruptors: From Endocrine to Metabolic Disruption."

42. California Department of Toxic Substances Control, "Flame Retardant Levels in California Breast Milk Decreasing," CDTSC (website), https://www.dtsc .ca.gov/SCP/PBDEsDecrease.cfm.

43. Greenpeace East Asia, *A Little Story About the Monsters in Your Closet* (Beijing: Greenpeace East Asia, 2014), http://www .greenpeace.org/eastasia/Global/eastasia /publications/reports/toxics/2013/A %20Little%20Story%20About%20the %20Monsters%20In%20Your%20Closet %20-%20Report.pdf.

44. National Academy Press, *Toxicological Risks of Selected Flame-Retardant Chemicals* (Washington, DC: NAP, 200), 213, https:// www.nap.edu/read/9841/chapter/12#231.

Index

Note: Page numbers in *italics* refer to figures, photos, and illustrations.

About the Authors

REBECCA BURGESS, M.Ed., is the executive director of Fibershed, the chair of the board for Carbon Cycle Institute, and the author of *Harvesting Color*. She is a vocationally trained weaver and natural dyer. She has over a decade of experience writing and implementing hands-on curricula that focus on the intersection of restoration ecology and fiber systems.  Burgess has built an extensive network of farmers and artisans in the Northern California Fibershed to pilot an innovative fiber systems model at the community scale. Her project has become internationally recognized with over fifty-three Fibershed communities now in existence.

A former archaeologist and Sierra Club activist, **COURTNEY WHITE** dropped out of the "conflict industry" in 1997 to cofound the Quivira Coalition, a nonprofit conservation organization dedicated to building a radical center among ranchers, conservationists, public land managers, scientists, and others around practices that improve economic and ecological resilience in western working landscapes.

In 2005 Wendell Berry included White's essay "The Working Wilderness" in his collection *The Way of Ignorance*. White is the author of *Revolution on the Range*; *Grass, Soil, Hope*; *The Age of Consequences*; and *Two Percent Solutions for the Planet*. He is also the author of *The Sun*, a mystery novel set on a working cattle ranch in northern New Mexico.

He lives in Santa Fe.